Disaster and Development

Disaster and Development

Editor

Sanjay Rohilla

scitus
academics

Disaster and Development

Edited by **Sanjay Rohilla**

Printed in 2017

ISBN: 978-1-68117-124-1

Library of Congress Control Number: 2015936546

SCITUS Academics LLC,
616, Corporate Way, Suite 2, 4766,
Valley Cottage, NY 10989

www.scitusacademics.com

Contents

Preface

The development process does not necessarily reduce vulnerability to natural hazards. Instead, it can unwittingly create new forms of vulnerability or exacerbate existing ones, impeding efforts to reduce poverty and promote growth, sometimes with tragic consequences. 'Win-win' solutions for securing sustainable development, reducing poverty and strengthening hazard resilience, therefore, need to be explicitly and actively sought, particularly as climate change looks set to increase the incidence of droughts and floods and the intensity of windstorms. Such solutions are best derived by integrating disaster risk reduction strategies and measures within the overall development framework, viewing disaster risk reduction as an integral component of the development process rather than as an end in its own right.

Editor

Chapter 1

Traditional Coping Strategies and Disaster Response: Examples from the South Pacific Region

Stephanie M. Fletcher[1], Jodi Thiessen[1], Anna Gero[2],
Michele Rumsey[1], Natasha Kuruppu[2],
and Juliet Willetts[2]

[1]World Health Organization Collaborating Centre for Nursing, Midwifery and Health Development, University of Technology, Sydney, Broadway, NSW 2007, Australia

[2]Institute for Sustainable Futures, University of Technology, Sydney, Broadway, NSW 2007, Australia

ABSTRACT

The Pacific Islands are vulnerable to climate change and increased risk of disasters not only because of their isolated and often low lying geographical setting but because of their economic status which renders them reliant on donor support. In a qualitative study exploring the adaptive capacity of Pacific Island Countries (PICs) across four countries, Cook Islands, Fiji, Samoa, and Vanuatu, it was clear that traditional coping strategies are consistently being applied as part of response to disasters and climate changes. This paper describes five common strategies employed in PICs as understood through this research: recognition of traditional methods; faith and religious beliefs; traditional governance and leadership; family and community involvement; and agriculture and food security. While this study does not trial the efficacy of these methods, it provides an indication of what methods are being used and therefore a starting point for further research into which of these traditional strategies are beneficial. These findings also provide important impetus for Pacific Island governments to recognise traditional approaches in their disaster preparedness and response processes.

INTRODUCTION

There exists evidence that global warming is contributing to increased numbers of observed extreme weather events, such as heat waves, extremes in precipitation, and the potential for increased number of severe storms [1]. For the Pacific, projected frequency of tropical cyclones may follow global trends of less frequent tropical cyclones and a likely increase in the relative proportion of severe storms by the end of the 21st century [2]. Coping with climate and weather extremes is a natural part of Pacific Islander's way of life. For instance, the 1997-98 El Nino events were blamed for widespread drought and famine in some Pacific islands, including major agricultural losses in Fiji, and widespread famine in parts of Papua New Guinea [3, 4]. Tropical cyclones, increased rainfall, flooding, and cyclone induced storm surges have resulted in extensive infrastructural and financial losses; particularly in Samoa, Cook Islands, and Fiji [2, 5–7]. It is anticipated that with projected climate changes, patterns of extreme weather including tropical cyclones may change, with an increased proportion of severe storms [4]. The extensive coastal boundaries, the numerous islands and atolls, long coastlines, and the density of population in coastal areas of South Pacific countries have made tsunamis, floods, and tidal abnormalities the most destructive problems in the region [8, 9].

In the context of this paper, "climate change" refers to any change in climate over time, whether due to natural variability or as a result of human activity [10]. A UNICEF report in 2010 indicates that climate change impacts in the Asia and the Pacific region will result in mortality and injury from extreme weather events; increase in risk of water scarcity; changes in the transmission, incidence, and distribution of water-, food-, and vector-borne diseases; migration and declining livelihoods; and heightened risk of food insecurity and child malnutrition [11]. Moreover, in the Pacific Island Countries (PICs) context, there are several "nonclimate" related factors and processes which affect the ability of communities to cope with climate stresses. These include (but are not limited to) rapid population growth, land ownership and kinship issues, intrusion of western culture into traditional governance, the erosion of traditional knowledge impacting agricultural practices, natural resource degradation, and increasing reliance on imports [12, 13]. Additionally, the small population size and economic vulnerability exacerbated by frequent disasters have resulted in the need for humanitarian assistance and development aid to manage the effects of disasters and climate change across the region [14, 15]. Both Australian government, local, regional, and international nongovernmental organizations (NGOs) and multilateral organizations including United Nations agencies are key development partners and provide rapid humanitarian assistance in PICs [15–19]. Several of these organizations also work to ensure the disaster response process transitions to postdisaster development programs to enhance overall in-country capacity [6, 20–22]. Disaster preparedness, risk reduction, and climate change adaptation are some of the thematic areas of development assistance in PICs [6, 21–24].

Historically, however, Pacific populations have applied their own traditional and local knowledge to respond to environmental challenges and changes [25]. These local and traditional knowledge and skills are increasingly being recognised as vital resources for adaptation [26]. Understanding local strategies used to prepare for, respond to, and recover from climate related disasters will help to define the Pacific Islands capacity to adapt to disasters and climate related events [25, 27]. This approach has being widely recognised and adopted within the field of climate change adaptation across developing countries as a part of response [28–30]. Both international and local leaders across the Pacific recognise the importance of traditional knowledge and coping skills to combat the effects of a changing and variable climate [14, 23, 30–33]. The Pacific Islands Framework for Action on Climate Change (PIFACC) [23] recognises and promotes the understanding and application of traditional knowledge to inform adaptation measures. To this end there has been a call for more research to link historical and anecdotal evidence of local coping strategies and for the integration of traditional knowledge into information management systems [23, 31]. Efforts have been made to capitalise on existing knowledge of regional cultural approaches and experiences with climate variability through consultative and research approaches [31]. It is therefore important for national disaster organizations and donors to recognise and incorporate traditional knowledge and skills in relevant planning and programming.

The purpose of this paper is to explore traditional coping strategies that may enhance long-term adaptive capacity to a variable and changing climate by looking specifically at examples from Pacific Islanders in their response to impacts of climate change and associated disasters. It draws on findings from a larger study which examined the adaptive capacity of the disaster response systems in four PICs to meet immediate humanitarian needs following a disaster [27]. This paper details the findings from the larger study giving examples of traditional coping strategies being used to address climate change impacts, including disasters, in terms of recognition of traditional coping strategies; faith and religious beliefs; traditional governance and leadership; family and community involvement; and agriculture and food security. These findings are useful for donor organizations and nongovernmental organizations to inform design and implementation of culturally relevant disaster risk reduction and climate change adaptation programs, in PICs and similar small island developing states.

METHODS

A qualitative research methodology was used including desktop review of the literature, policy analysis, and ninety interviews with high level professionals involved in disaster response from Australian and PIC government ministries, local and regional NGOs, and international humanitarian and donor organizations. The research process was guided by a Project Reference Group (PRG) comprising key experts from Australia and the Pacific as a form of structured stakeholder oversight and engagement in the research. Furthermore, a Conceptual Framework [27] was developed to provide the scope for this study. This framework described a cycle of adaptive learning within which the adaptive capacity of the "disaster response system" (DRS) is influenced by a range of key determinants [34]. Specific determinants of adaptive capacity were used to assess the DRS and drew upon literature spanning Earth System Governance [35], climate change adaptation [34], health resources [36–38], resilience in institutions [39], and practice theory [40]. Key determinants of adaptive capacity were identified as being interorganizational including architecture; agency; adaptiveness [35, 41]; intraorganizational objective including leadership, management, and governance structures [42]; technical capacity, tools, methods, and approaches [43]; human resource for health governance and management systems [36, 37]; and finally intra-organizational subjective including risk perceptions [34]; self-efficacy beliefs [35, 39]; elements of social practice [40]; and traditional coping strategies [40, 44]. The subjective determinant of "traditional coping strategies" is the focus of this paper and therefore expands specifically on findings regarding this key determinant [27].

A semistructured interview guide informed by a conceptual framework was used to guide face-to-face and telephone interviews with all 90 participants; and probing conducted as needed. The interview style, developed with the guidance of the PRG, was designed to appreciate cultural differences including the need for clear questions and flexibility when interviewing people from various cultural backgrounds. Relevant organizations were identified with the assistance of the PRG, with additional organizations identified through a snow ball sampling technique (Atkinson and Flint, 2001). The interviews were conducted in both Australia and four Pacific Island Countries in May–July 2012 along with workshops with in-country stakeholders to test the preliminary analysis of interviews.

Four PICs were investigated in detail to provide more of a regional viewpoint: Cook Islands, Fiji, Samoa, and Vanuatu. These countries were selected with the aim to include a variety of geographical settings (including Melanesia and Polynesia); varied policy landscapes (regarding climate change, disaster, and health-related policies and plans); mix of countries with recent significant climate-driven disasters and those in the more distant past (relating to degrees of memory and levels of preparedness); and countries that experience tropical cyclones as an example of a rapid onset disaster. The scope of the study was limited to health related humanitarian needs and as such researchers were interested in obtaining a mix of countries including high and low HRH density. Development indicators (e.g., varied Human Development Index, see UNDP, 2011) were also used as a criterion for selection to include countries at different stages of development. A full account of the methodology can be found in [27].

RESULTS AND DISCUSSIONS

Historically, the study of the traditional knowledge was the purview of anthropologists and sociologists. However, other disciplines have come to recognise the importance of the broader application and use of this knowledge before it disappears; and the value of the application of traditional methods to cope with environmental hazards is now being widely recognised [33, 45]. External influences from western colonisation, governance reforms, and development assistance have contributed to some loss of traditional and local knowledge and coping strategies [25]. Increasing commercialisation and urbanisation and the subsequent erosion of traditional social networks have also contributed to the loss of traditional knowledge and practices [13]. Within a generation traditional pass times centring on the extended family or local community have been replaced by individualised and home-based recreation [13]. Historically, livelihoods were constructed and maintained in a way that accounted for climate stress based on generations of experience of living with environmental uncertainty [12].

Members of the Pacific community recognise that with time, traditional knowledge is eroding due to a combination of factors such as migration, urbanization, and the passing away of elders [32]. It is essential therefore that beneficial traditional knowledge and cultural values are used and preserved in order to improve the ability of Pacific Islanders to build long-term resilience to natural disasters over time [32].

It became evident throughout the interviews conducted in PICs that traditional coping strategies are an important part of life and where some have been eroded, in particular with agriculture and food security, there is a current move back to integrating traditional methods, to improve disaster preparedness and response. The following section provides examples from the respondents as well as from the literature, outlining Pacific Islanders traditional coping strategies for disaster response and climate change adaptation. Table 1 summarises findings around traditional knowledge and coping strategies from the Pacific Region. However, it is recognised that continuing cultural and demographic change across the region is likely to influence traditional disaster-related coping mechanisms, including those mentioned above.

Table 1: Summary of traditional coping strategies to disasters and climate changes

Key areas	Summary of traditional knowledge and coping strategies for response to disasters and climate change in Pacific Island Countries
Recognition of traditional coping strategies	(i) Secretariat of the Pacific Community's (SPC) list of crops from the climate ready collection. (ii) Glossaries of traditional climate change terms in some Pacific countries. (iii) Pacific Red Cross societies Toolkit to assess natural disasters vulnerability and response capacity.
Faith and Religious beliefs	(i) Faith-based systems and their interventions are entrenched in the social system and can build resilience. (ii) Recognition of churches, their role in community life, including use as emergency centres or disaster shelters; provision of postdisaster counselling. (iii) Churches may be a source of resources, volunteers, welfare programs for the poor, and needy including non-members.

Traditional governance and leadership	(i) Indication of some integration of political and traditional governance systems for disaster management. (ii) Active involvement of traditional governance structures in village life, disaster management, and climate change adaptation. (iii) Use of traditional leadership networks to share information and communicate in times of disasters. (iv) Involvement of schools, churches, and the community in disaster preparedness and response.
Family and Community Involvement	(i) Extended family system and kinship ties provide a critical support structures in times of disaster. (ii) Movement of families from high risk areas to less vulnerable areas inland during disasters. (iii) Fostering of relationships improves coordination of response teams and helps develop good coping skills.
Agriculture and food security	(i) Traditional agricultural disaster-preparation techniques to preserve seedlings and seeds. (ii) Documenting food preservation techniques and consideration of inclusion in school curricula. (iii) Integrated farming system approach that incorporates the planting of fruit trees with root crops. (iv) Introducing less common foods and reintroducing some forgotten foods from traditional diets. (v) Using organic agricultural practices-pesticide free and no chemicals.

RECOGNITION OF TRADITIONAL COPING STRATEGIES

Disaster response stakeholders in four Pacific Island Countries indicated that traditional coping strategies were important to help communities cope with the effects of disasters and, increasingly, the impacts of climate change. The importance of recognising and applying traditional knowledge and coping strategies was widely recognised by interviewees. A NGO representative from Vanuatu summed it up succinctly.

"We are adaptable and will find new ways to integrate science and traditional knowledge. There are traditional skills–food security and local capacity."

Government, donor and NGO communities across all countries agreed that there was "a lot of capacity that can be tapped into the traditional cultures .... there is great potential here because of strong communities in place."

Unfortunately, traditional strategies were found to not always be recognised, relied on, and incorporated in national disaster response mechanisms. This may be due to the fact that "the traditional coping methods documentation is not there, it's oral." According to one regional health respondent in Vanuatu,

"There are strong partnerships between islands. This includes strong kinship ties. People develop good coping skills and foster relationships.... that are very important even during coordination of teams. We have local mechanisms in place that we (at the national level) seem to overlook. Legally, these traditional structures are not incorporated. It needs to become part of the local ways of responding."

Also a survey has indicated that the use of traditional medicine is not often acknowledged because of its association with "witchcraft" [46] (cited by WHO 2001). However, a health sector respondent indicated progress had been made with the successful publication of the traditional medicine book, Na Wai Vakaviti [47], by the Women's Association for Natural Medicinal Therapy, (Wainimate) Fiji. The respondent went on to say that "we included the knowledge and the practice from those who have been practicing; they documented it and some those [traditional medicinal therapies] have been tested in the science lab in the University of South Pacific."

In this study, traditional practices were chosen for illustration based on their frequency of use or discussion by responders and based on first hand experiences gained by the researchers whilst working in the region. According to WHO, there is increasing recognition amongst PICs for the need to regulate the use of traditional medicines [48, 49], although for the most part this has not yet occurred. This is due to the fact that harmful and adverse reactions can result from the use of poor quality or inappropriately administered traditional medicines and practices [49, 50]. In recognition of the importance of the use of safe and effective traditional medicinal practices and products as an important form of health care for many people in the Western Pacific Region and to maximize the potential contribution of traditional medicine to primary health care, the WHO has developed the Regional Strategy for Traditional Medicine in the Western Pacific, as a guide to relevant stakeholders [49]. However, there is the need for more empirical evidence to determine the efficacy of traditional medicine practices and products, their use in conjunction with contemporary medicine, and the need for adequate information to guide consumers to make informed decisions [49].

In recognition of the strong traditional systems in the Pacific, the Climate Change and Development (CCD) Community, of the online Pacific Solution Exchange, recently held an online discussion that called for the incorporation of traditional knowledge into national planning in order to address climate change impacts and disaster risk reduction and to balance modern technology and science with such traditional practices in the PICs. According to a member of the CCD community, "there is keen interest among both older and younger generations of islanders to record this traditional knowledge and to integrate it with scientific approaches" [32].

Advancements have already been made in the preservation of traditional knowledge. For instance, the Climate Change and Development Community reported some examples which includes list of crops from the Secretariat of the Pacific Community's (SPC) climate ready collection and glossaries of traditional climate change terms in some Pacific countries, such as the iTaukei Glossary in Fiji. There was also a call for the inclusion of traditional knowledge in school curriculums, especially in remote areas with limited or no access to the internet. There was acknowledgement by several interviewees of the need to embrace traditional knowledge as an important and essential tool for disaster risk management and to combat climate change. This will require the combination of traditional knowledge with scientific knowledge, in order to promote the environmental, economic and social sustainability necessary when working with local communities who will be the primary beneficiaries of the products. To achieve this union, careful assessment and analysis are required to ensure that innate difficulties of blending traditional knowledge with scientific knowledge are addressed and that traditional value systems are upheld [32]. For instance, the Red Cross has introduced a toolkit for Pacific Red Cross societies and island communities to assess their vulnerability and capacity in times of natural disasters. Several techniques to raise community awareness were included in the toolkit, including [51] the following:

- recording of seasonal patterns over time,
- documenting social changes over time that impact lives and conditions, such as local hazards, construction material and design, vegetation, livestock, and environmental changes,
- mapping of risks, hazards, and disasters response resources,
- sharing of knowledge between communities,
- formal approval for traditional knowledge and disaster preparedness to be included in primary school curricula, planning, and social development programs,
- effective risk and hazard communication between communities and local authorities.

It must be noted, however, that, in the quest to blend science and traditional knowledge, care should be taken to ensure that practices that have a deleterious effect on populations are not reinforced. This is because not all traditional coping strategies are beneficial and some, in fact, can be harmful and infringe on human rights [52]. One such example was cited by a respondent in Samoa. After the tsunami in 2009, some first responders to the affected area prioritised the removal of the dead bodies before the injured, because it was a culturally acceptable way to pay respect to the dead. Prioritising the dead over the injured, thus delaying treatment, could result in deterioration of their health status. There is limited published information about the incidence and impact of some harmful traditional and cultural practices, and in many cases, these practices may not be seen as being harmful in their culture. Engaging with local communities to determine their attitudes to certain practices as well as to undertake further studies on the incidence, causes and consequences of harmful practices are recommended.

FAITH AND RELIGIOUS BELIEFS

Pacific Islands are well known for their strong religious ties and the involvement of religious groups in the development and leadership of communities. This study found that faith (predominantly Christianity) and involvement in religious communities were an important coping strategy across the case study countries. In fact, the Council of Churches or equivalent organization in each country was an important social and cultural institution and active in disasters response. Churches were found to play an important role in community life and often used as emergency centres or shelters in times of disaster. This church-based respondent described their community disaster evacuation plan:

"Everyone is instructed that once they hear the warning they should come to the shelter (church building) with their emergency backpacks [which are already prepared with emergency supplies to last 48 hours] and walk up to the [designated] Primary School."

The role of the church is widely recognised across all levels of society. According to one NGO representative in Vanuatu, "women's church groups are the strongest community group." The role of the church in posttsunami response in Samoa was often highlighted by respondents. This included the role of the Church as a key agent in the provision of counselling and psychosocial support to survivors and affected families and response personnel, bringing comfort to the people and attending to the spiritual needs of community and response personnel alike. It was evident that the Church was "held in high esteem in Samoan society." According to one church representative,

"at the time of the tsunami in 2009, our previous chairman (deceased) was immediately contacted by the Minister for Health for comfort. The Chairman went to the main hospital and was there during the entire day comforting the health staff and the nation at large. His prayers and speeches were broadcasted live across the nation throughout the day."

International and local humanitarian organizations recognise the benefits of working with churches as a powerful social agent, to advance disaster risk reduction and climate change adaptation in PICs [53, 54]. An example of the role of the church community was given by a Samoa respondent who indicated that one church in particular was the first to be on the ground when the tsunami hit. This was possible because they had all the necessary resources in terms of relief supplies and funds. They were able to immediately mobilise land and air transportation to distribute relief items, even before the Red Cross and Government representatives came in to do their assessments. They were also reported to able to provide shelter to affected communities as well as address psychosocial needs. These comments were corroborated by other Samoan respondents.

The importance of faith to build resilience was summed up by one health respondent:

"Faith based systems and their interventions were already entrenched in the social system so people were able to turn to these during the tsunami. This is why people were more resilient than expected; they were able to recover quickly because of these social groups."

However, the role of the church in providing psychosocial and mental health support was also viewed as limited by some respondents. One regional representative in Fiji indicated that "I do not think the churches have the skills or the capacity to do that..... and the churches will only speak to the Christian population. So the other people are left out-the non-Christians." Other respondents indicated a need to have a system in place to identify persons who are isolated from these support structures such as ethnic minorities, disabled, and those who cannot go to or are not members of a local church. The extent to which aid is exclusively provided to Church members was not determined by this study, but there was no indication that this was common practice in all case study countries. Neither was it determined to what degree Church counselling was effective. However, regardless of its impact and effectiveness, Pacific Islanders see faith-based interventions as part of their lives and culturally appropriate. It is against this background that Taylor (page 177, 2003) recommended that the "power of religious belief and of social justice in the process of healing" is a concept that should be carefully considered by overseas responders when working in the Pacific. [55]. This study therefore recommends that further research be conducted and that opportunities for knowledge sharing be encouraged, in order to better understand the effectiveness and importance of faith-based interventions.

TRADITIONAL GOVERNANCE AND LEADERSHIP

Strong traditional local governance structures are common across Pacific Island Countries, examples of which include formal Women's Committees and Council of Chiefs [53, 56, 57]. The integration of political and traditional governance has been credited for the relative stability of governance, especially in Polynesian countries [57]. Traditional governance structures are an active part of village life and play a significant role in disaster management and climate change adaptation [56, 58]. In some cases, like Samoa, their links to the national government are formalised through hierarchical traditional governance structure [56]. Understanding and incorporating the roles and essential functions of these hierarchical structures in disaster risk reduction and climate change adaptation programs can be an essential asset [53].

In Vanuatu, for example, it was reported that communities look to the Council of Chiefs for instruction and leadership in times of disasters. A health sector respondent reported that "they [Chiefs] use a sea shell to indicate a meeting or to prepare for a Cyclone or disaster. They also ring a bell–a different type of ringing represents different type of warnings." Village Chiefs, who also serve as chairmen of the Area Councils (a form of local level of governance in Vanuatu), use their traditional networks in communities to share information in times of disasters. Similarly in Samoa, the Village Council is actively involved in the enforcement of adaptation activities at the community level and integration of disaster risk reduction and climate change adaptation. Evidence of this has been described through the Ministry of Women Community and Social Development's (MWCSD) Aiga ma Nuu Manuia Programme (Family Wellbeing Programme), which has gained recognition in Samoa [59]. Village Leaders reportedly ensure that "the bell is rung to warn of impending danger such as a tsunami." The Women's Committees at village level are also key institutions for communities both during and after disasters. A government official in Samoa reported that

"In times of disasters, The MWCSD mobilised these agents (village leaders and women's committees) at the local levels to ensure relief reaches the appropriate persons/household. They also feed into the national government emergency operations centre. The Women's groups help out at the disaster shelters cooking, feeding, counting wards."

Coordination for disaster preparedness at the village level in Samoa was reported to involve regular meetings with schools, churches, and the community to ensure that everyone is involved during disaster response. The role of traditional leadership and the active involvement of women's groups in local governance cannot be underestimated or ignored. Their involvement in the planning and implementation of disaster management and climate change adaptation programs will ensure that local knowledge and culturally appropriate approaches are applied and that there is community buy-in and support for adaptation activities.

FAMILY AND COMMUNITY INVOLVEMENT

The role of strong family ties and support systems as a strategy for coping with disasters was also identified. Respondents indicated that a strong extended family system was critical for support in times of disaster. In Fiji a government official noted that

"people are strong because of the support from the family members; so when we are here in Suva and disaster strikes somewhere else, we know exactly where our relatives are. So then we can mobilise the families here to do something and then we take it to them there. Even though the Government is going to everyone, we go and see our individual families to encourage and support them."

A NGO representative in Vanuatu summed it up as such; "I'm surprised at the natural resilience-they[communities] can come together at the spur of the moment when they need to."

Another example of the community coming together was described by respondents in Samoa. The Samoa Volunteer Emergency Response

Team (VERT) and Water Safety Project (WSP) set up to strengthen emergency response capacity of the Samoa Fire and Emergency Services Authority (SFESA), the first responder to disasters and emergencies that requires immediate evacuation, search and rescue, first aid, and medical evacuation [60]. While in its early, capacity-building stages, the VERT-WSP programme aims to minimise dependence on permanent staff required to man base stations in the event of concurrent emergencies [60]. According to one respondent "Post tsunami, the local farmers and carpenters [who were involved in the VERT] were some of the best actors in the response, because they were used to working in rough environments and this was an asset to search and rescue operations."

Other social support systems include the contribution of the community to the immediate needs of the affected, a common practice in PICs. In Fiji, for example, when there is an appeal for disaster relief, people donate whatever they have "school children sometimes donate clothing and one dollar each." This allows the government to focus on "funds for building the homes and supplying of building materials." Another example was given from Samoa, where the fact that some families own both on the coastal lands and inland plantations, facilitating moving of people from hazard zones on the coast to safer areas inland.

AGRICULTURE AND FOOD SECURITY

Pacific communities have traditionally faced food security issues arising from extended dry periods associated with El Niño/La Niña-Southern Oscillation (ENSO) events or from the destruction of agricultural produce by cyclones and floods and over time have developed ways to preserve food to cope with such situations [61, 62] Climate change may bring about an exacerbation of such extremes in the Pacific [7]. Several respondents alluded to the fact that there was a need to return to "traditional ways of preserving food to withstand during times of crisis" with the added threats climate change may bring. However, one NGO respondent went on to say that although documentation of traditional agricultural practices is limited, "it has been passed down from generations [so] they know what to do."

There has been a call from the Climate Change and Development Community (CCDC) to renew traditional knowledge of agricultural and farming practices. In some instances, they suggested that an ideal approach could be an "integrated farming system approach" that incorporates the planting of fruit trees with root crops [32]. Examples of an integrated farming approach were cited in Samoa where it was reported that current efforts by local government organizations and NGOs are "seeking to include traditional agriculture and horticulture for food security and livelihoods through economic programs." These include how to plant, what are the best sites, organic methods, and how to cook the produce and the importance of traditional ways of growing. For example, one NGO respondent suggested that "having a backup plantation for cyclone period by growing taro patches, since bananas and breadfruit trees fall down." Existing documentation of other traditional agricultural disaster-preparation techniques come from the Pacific Climate Change and Development Community such as cutting the stems of ready to harvest cassava plants to save them from damage during a cyclone; burying the harvest of taros, so it lasts for up to six months; or preserving breadfruit and pandanus by making them into a paste in Kiribati [32]. Also, "in the old days," after a cyclone Samoan communities would salvage ready to harvest root-crop tubers, filled with valuable starch, which were then preserved in underground fermentation rooms [32].

One respondent in Fiji indicated that "knowing that climate change will bring about risks to food we have to do things to suit us and food preservation is important." To this end, it was reported that "a school teacher is preparing documents for the food security." However, the respondent went on to say that there was a need for funding to produce the documents, but there was limited funding for the food security program.

Another approach suggested by some respondents was to introduce less common foods and reintroduce some forgotten foods into diets. According to one respondent, "Cassava was traditionally perceived as pig food, but efforts are being made to re-introduce it in the Samoan diet." In Fiji, one respondent indicated that there was a risk for overfishing, and suggested alternatives. Traditionally, tuna and Skipjack are not eaten in Fiji [63], but with changing behaviour and adapting to the overfished resources, these species may become a viable alternative in Fijian diets.

Using organic agricultural practices is one way to preserve the produce. Both Fijian and Samoa respondents mentioned that the governments were supporting the growth of pesticide free and chemical free crops. One Samoan respondent indicated that while the international market wants large fruits, those that are found in the local markets are small and pesticides-free. Another regional respondent concurred; countries with strong tourism sectors such as Fiji, and increasingly Samoa and Cook Islands, have major markets for organically grown vegetables.

However, some harmful practices were identified as compounding the negative impacts of climate change. These include the use of dynamite for fishing, use of burning to clear vegetation for planting increasing the risk of soil erosion, landslides, and flooding. Similar sentiments were expressed by members of the CCDC, citing examples of maladaptation [32]. Examples of how misunderstanding of local traditional systems and practices can undermine community resilience and adaptation in times of disaster was also described. For instance, in Samoa according to one government official, after the tsunami in 2009 some NGOs were offering "cash-for-work." When the local authorities asked for the community's assistance later, community members requested payment. The respondent found this very unusual as previously community spirit was the default way of response. The respondent perceived this as affecting the resilience and mindset of villagers who were accustomed to working to build their own community. Furthermore, offering cash-for-work is not sustainable by the local authorities and so may have affected the adaptation capacity of the community. This is one example of how understanding the local value system is useful and how misunderstanding of such value systems may be negatively impacted by inappropriate external influences.

LIMITATIONS AND FURTHER RESEARCH

As a qualitative methodology was used, this study cannot report on the efficacy of the traditional coping strategies described. As such, these findings must be interpreted in that context. However, the rigorous methodology ensured inclusivity of a very large sample size of various sectors and stakeholders from the region, supported by literature and peer-review by subject experts from the region; providing triangulation of the broad thematic areas. Moreover the representation of Pacific Islanders on the PRG who played a critical role in informing and guiding the interpretation and analysis of research outcomes provide further validation of the findings.

Exploring a mix of countries allowed a more regional view of how disaster response plays out in different settings which proved a strength of the study. However, in terms of traditional coping strategies this broad viewpoint was also a limitation. Pacific Islands are unique in their culture and traditions, and as noted by Nunn (2009), the interpretation of the climate change concept is influenced by local culture and religion. As such traditional coping strategies that work in one country may not necessarily work the same way in another country, even across countries within the same region. As this study focussed on a regional overview, the traditional mechanisms described may be particular for the respondents' country; however, the broad thematic areas were common across all countries. Specific traditional practices for the complex and varying cultures of the Pacific would need to be explored on a very local level. Therefore further research into some of these strategies and their effectiveness in building community resilience throughout the disaster continuum is recommended.

CONCLUSIONS

This paper has provided examples of ways in which traditional practices and coping strategies for disaster response are drawn upon in the Pacific. The use of stakeholder interviews with key members of disaster response organizations in Australia and the Pacific allowed specific insight into the ways in which Pacific islanders apply traditional knowledge and coping strategies to prepare for and respond to disasters. The respondents in this study provided many examples of traditional coping strategies being used to address climate change impacts. The most important themes that emerged include recognition of traditional coping strategies; faith and religious beliefs; traditional governance and leadership; family and community involvement; and agriculture and food security. Acknowledging and preserving these useful traditional coping strategies can contribute significantly in helping the donor and NGO community in integrating them into culturally relevant programming. Furthermore they can also provide important impetus for Pacific Island governments to embrace these approaches in their disaster preparedness and response processes. While evidence of effectiveness is needed, traditional practices are still important mechanisms that are naturally integrated into the lives of Pacific Islanders as described by respondents to this study. Therefore it is important that traditional ways of operating and supporting local communities are understood by external organizations seeking to offer support and provide humanitarian response in Pacific Island Countries. Understanding the historically embedded social support systems that encourage self-help and working together is likely to enhance the outcomes of donor programs and development initiatives.

ACKNOWLEDGMENTS

This work was funded by the National Climate Change Adaptation Research Facility (NCCARF). The authors gratefully acknowledge this funding. The views expressed in this paper are those of the authors and not necessarily those of NCCARF. The authors thank all research participants for their contributions, including their Project Reference Group. This work was guided by the principles of ethics and was approved by the Human Research Ethics Committee at UTS.

REFERENCES

1. D. Coumou and S. Rahmstorf, "A decade of weather extremes," Nature Climate Change, vol. 2, no. 7, pp. 491–496, 2012. ·
2. T. R. Knutson, J. L. McBride, J. Chan et al., "Tropical cyclones and climate change," Nature Geoscience, vol. 3, no. 3, pp. 157–163, 2010. · ·
3. W. Kininmonth, "The 1997-1998 El Nino event: a scientific and technical retrospective," WMO Report, 1999.
4. Australian Bureau of Meteorology and CSIRO, "Pacific Climate Change Science Program (PCCSP): current and future climate of Pacific Island countries," International Climate Change Adaptation Initiative (ICCAI), Collaboration between Pacific

Osland Countries, Australian Bureau of Meteorology and Commonwealth Scientific and Industrial Research Organization, 2011.

5. A. Olsthoorn, W. Maunder, and R. S. Tol, "Tropical cyclones in the Southwest Pacific: impacts on Pacific Island countries with particular reference to Fiji," in Climate, Change and Risk, T. Downing, A. Olsthoorn, and R. Tol, Eds., pp. 221–244, Routledge, 2002.
6. J. Campbell, "Consolidating mutual assistance in disaster management within the Pacific: principles and application," in Proceedings of the 7th South Pacific Regional IDNDR Disaster Management Meeting, September, 1998.
7. K. J. E. Walsh, K. L. McInnes, and J. L. McBride, "Climate change impacts on tropical cyclones and extreme sea levels in the South Pacific—a regional assessment," Global and Planetary Change, vol. 80-81, pp. 149–164, 2012. · ·
8. A. J. McMichael, D. H. Campbell-Lendrum, C. F. Corvalán et al., "Climate change and human health: risks and responses," Tech. Rep., WHO Library, Geneva, Switzerland, 2003.
9. ADRC, Natural Disasters Data Book 2009, 2009.
10. IPCC, "Summary for policymakers," Climate Change 2007: the Physical Science Basis, Contribution of Working Group I to the Fourth Assessment Report, Cambridge University Press, Cambridge, UK, 2007.
11. M. Urbano, N. Maclellan, T. Ruff, and G. Blashki, "Climate change and children in the Pacific Islands,"UNICEF Pacific, 2010.
12. O. Warrick, "Climate change and social change: vulnerability and adaptation in rural Vanuatu," The University of Waikato, Hamilton, New Zealand, 2010.
13. M. Pelling and J. I. Uitto, "Small island developing states: natural disaster vulnerability and global change," Global Environmental Change: Part B, vol. 3, no. 2, pp. 49–62, 2001. · ·
14. S. Fletcher, A. Gero, M. Willetts et al., "Review of Australia›s overseas disaster and emergency response sector," Prepared for NCCARF, WHO Collaborating Centre and the Institute for Sustainable Futures, University of Technology, Sydney, Australia, 2012.

15. N. Maclellan, "Turning the tide: improving access to climate financing in the Pacific Islands," Lowy Institute for International Policy, 2011.
16. Australian Government, "Humanitarian preparedness and response," 2013,http://aid.dfat.gov.au/makediff/humanitarian/Pages/default.aspx.
17. Australian Government, "Regional Pacific," 2013,http://aid.dfat.gov.au/countries/pacific/rp/Pages/home.aspx.
18. CARE, "Care humanitarian action," 2011, http://www.care.org.au/Document.Doc?id=558.
19. Caritas Australia, "Annual report 2010-2011," 2011.
20. AusAID, "Pacific regional aid strategy 2004–2009," Tech. Rep., Australian Agency for International Development (AusAID), Canberra, Australia, 2004.
21. AusAID, "Integration in practice-integrating disaster risk reduction, climate change and environmental considerations in AusAID programs," Tech. Rep., Australian Agency for International Development (AusAID), Canberra, Australia, 2010.
22. Australian Government, "International climate change adaptation initiative," 2011,http://aid.dfat.gov.au/aidissues/climatechange/Pages/adaptation_initiative.aspx.
23. SPREP, "Pacific Islands Framework for Action on Climate Change (PIFACC) 2006–2015," Tech. Rep., Apia, Samoa, 2007.
24. SOPAC, "Pacific disaster risk reduction and disaster management: a framework for action 2005–2015," Tech. Rep., SOPAC, Suva, Fiji Islands, 2010.
25. I. Kelman, "Hearing local voices from small island developing states for climate change," Local Environment, vol. 15, no. 7, pp. 605–619, 2010.
26. L. A. Nurse, G. Sem, J. E. Hay et al., "Small island states," Climate Change 2001: Impacts, Adaptationand Vulnerability, 2001.
27. A. Gero, S. Fletcher, M. Rumsey et al., "Disaster response and climate change in the Pacific," Tech. Rep., National Climate Change Adaptation Research Facility (NCCARF), Sydney, Australia, 2013.
28. M. K. van Aalst, T. Cannon, and I. Burton, "Community level adaptation to climate change: the potential role of participatory

community risk assessment," Global Environmental Change, vol. 18, no. 1, pp. 165–179, 2008. · ·

29. M. S. Reed, A. J. Dougill, and M. J. Taylor, "Integrating local and scientific knowledge for adaptation to land degradation: kalahari rangeland management options," Land Degradation and Development, vol. 18, no. 3, pp. 249–268, 2007. · ·
30. A. Gero, K. Méheux, and D. Dominey-Howes, "Integrating community based disaster risk reduction and climate change adaptation: examples from the Pacific," Natural Hazards and Earth System Science, vol. 11, no. 1, pp. 101–113, 2011. · ·
31. J. Barnett and M. Busse, "Ethnographic perspectives on resilience to climate variability in Pacific Island countries," Tech. Rep., Asia-Pacific Network (APN) for Global Change Research, Projects, 2001.
32. J. Berrel and E. Philips, "Role of traditional knowledge (TK) to address climate change impacts," Tech. Rep., Climate Change & Development (CCD) Community, of the Pacific Solution Exchange, 2013.
33. Commonwealth of Australia, "Australia›s international development assistance program 2012-13," Tech. Rep., Country and Regional Programs, 2012.
34. M. Ekström, N. Kuruppu, R. L. Wilby et al., "Examination of climate risk using a modified uncertainty matrix framework—applications in the water sector," Global Environmental Change, vol. 23, no. 1, pp. 115–129, 2012. ·
35. F. Biermann, "‹Earth system governance› as a crosscutting theme of global change research," Global Environmental Change, vol. 17, no. 3-4, pp. 326–337, 2007. · ·
36. World Health Organization, "The world health report 2006—working together for health," Tech. Rep., World Health Organization, Geneva, Switzerland, 2006.
37. WHO/WPRO, "Human resources for health action framework for the Western Pacific Region: 2011–2015," Tech. Rep., World Health Organization Regional Office for the Western Pacific Region, Manila, Philippines, 2011.
38. K. J. Bowen, S. Friel, K. Ebi, C. D. Butler, F. Miller, and A. J. McMichael, "Governing for a healthy population: towards an

understanding of how decision-making will determine our global health in a changing climate," International Journal of Environmental Research and Public Health, vol. 9, no. 1, pp. 55–72, 2011. · ·

39. S. McManus, E. Seville, D. Brunsdon, and J. Vargo, "Resilience management: a framework for assessing and improving the resilience of organizations," Tech. Rep., Resilient Organizations Research Report, 2007.
40. Y. Strengers, "Conceptualising everyday practices: composition, reproduction and change," Tech. Rep. no. 6, Carbon Neutral Communities, 2010.
41. I. Kelman and J. J. West, "Climate change and small island developing states: a critical review,"Ecological and Environmental Anthropology, vol. 5, no. 1, pp. 1–16, 2009.
42. S. McManus, E. Seville, J. Vargo, and D. Brunsdon, "Facilitated process for improving organizational resilience," Natural Hazards Review, vol. 9, no. 2, pp. 81–90, 2008. · ·
43. Food Agricultural Organization (FAO), Rome, Italy, 2005.
44. J. Mercer, D. Dominey-Howes, I. Kelman, and K. Lloyd, "The potential for combining indigenous and western knowledge in reducing vulnerability to environmental hazards in small island developing states," Environmental Hazards, vol. 7, no. 4, pp. 245–256, 2007. · ·
45. J. Newton, "An assessment of coping with environmental hazards in northern Aboriginal communities,"Canadian Geographer, vol. 39, no. 2, pp. 112–120, 1995.
46. K. Chen, "Fiji national workshop on traditional medicine," Mission Report 25–31, Manila, Philippines, 2001.
47. R. Singh, "Book of healing," The Fiji Times Online, 2009.
48. World Health Organization (WHO), "Legal status of traditional medicine and complementary/alternative medicine: a worldwide review," 2001.
49. WHO/WPRO, "The regional strategy for rraditional medicine in the western pacific (2011–2020)," Tech. Rep., World Health Organization Regional Office for the Western Pacific Region, Manila, Philippines, 2012.

50. WHO, "How safe is traditional medicine," 2005, http://www.who.int/features/qa/20/en/index.html.

51. N. Maclellan, "Preparing for storms: protecting the islands," Islands Business International, 2011.

52. UN ESCAP, "Harmful traditional practices in three countries of South Asia: culture, human rights and violence against women," Gender and Development Discussion Paper Series 21, United Nations Economic and Social Commission for Asia and the Pacific (UN ESCAP), Bangkok, Thailand, 2007.

53. A. Gero, K. Méheux, and D. Dominey-Howes, "Integrating disaster risk reduction and climate change adaptation in the Pacific," Climate and Development, vol. 3, no. 4, pp. 310–327, 2011. ·

54. Pacific Conference of Churches, "Pacific Conference of Churches Strategic Plan 2008–2012," Suva, Fiji, 2007.

55. A. J. W. Taylor, "Cross-cultural interaction in the appraisal of disaster trauma in three Pacific Island countries," Asia Pacific Viewpoint, vol. 44, no. 2, pp. 177–193, 2003.

56. E. Huffer and A. So›o, "Beyond governance in Samoa: understanding Samoan political thought,"Contemporary Pacific, vol. 17, no. 2, pp. 311–333, 2005.

57. C. Saldanha, "Strategies for good governance in the Pacific," Asian-Pacific Economic Literature, vol. 18, no. 2, pp. 30–43, 2004. · ·

58. E. Huffer and A. So›o, "Consensus versus dissent: democracy, pluralism and governance in Samoa," Asia Pacific Viewpoint, vol. 44, no. 3, pp. 281–304, 2003.

59. G. Kenny, "Climate resilience in Samoa: capacity assessment and enhancement consultancy," Final Report, Apia, Samoa, 2012.

60. Samoa Fire and Emergency Services Authority, "Samoa volunteer emergency response team and water safety project," Project Design Document, Apia, Samoa, 2010.

61. N. Kuruppu and D. Liverman, "Mental preparation for climate adaptation: the role of cognition and culture in enhancing adaptive capacity of water management in Kiribati," Global Environmental Change, vol. 21, no. 2, pp. 657–669, 2011. · ·

62. J. Barr and P. Emergency, "Drought assessment: the 1997-98 El Niño drought in Papua New Guinea and the Solomon Islands," Australian Journal of Emergency Management, vol. 14, no. 2, pp. 35–41, 1999.

63. R. E. Kearney, "An assessment of the skipjack and baitfish resources of Fiji," 1984.

Chapter 2

Improvements of Volcanic Ash Fall Forecasts Issued by the Japan Meteorological Agency

Yoshihiko Hasegawa[1], Akira Sugai[1], Yosuke Hayashi[1], Yuta Hayashi[1], Shoji Saito+, and Toshiki Shimbori[2]

[1]Japan Meteorological Agency, Chiyoda-ku, Tokyo, Japan
[2]Meteorological Research Institute, Tsukuba, Ibaraki, Japan

ABSTRACT

Since March 2008, the Japan Meteorological Agency (JMA) has operated a Volcanic Ash Fall Forecast (VAFF) system to issue forecasts of areas in Japan where ash falls are expected following volcanic eruptions. The hazardous effects of ash falls vary according to the quantity of ash generated and have different consequences for agriculture, communication and transport networks, and buildings. The prediction techniques based on the JMA Regional Atmospheric Transport Model were recently revised by the Meteorological Research Institute of JMA to allow predictions of the quantity and areas of ash fall, and areas to be affected by lapilli fall. JMA plans further improvements to the VAFF system to address the needs of people living near active volcanoes and to take into account expert advice in fields such as volcanology, disaster prevention, and broadcasting and medical services. The improved VAFF system will provide three types of information: a regular information before possible eruption, a brief preliminary forecast issued 5–10 minutes after an eruption, and a full forecast issued 20–30 minutes after an eruption. JMA has developed a table to accompany VAFFs that categorizes the hazardous effects for people according to various quantities of ash fall, and provides advice on preparedness measures and actions to be taken when an ash fall occurs. The new VAFF system will also provide predictions of areas of expected lapilli fall. JMA will start operation of the new VAFF system in spring 2015.

INTRODUCTION

Since March 2008, the Japan Meteorological Agency (JMA) has operated the Volcanic Ash Fall Forecast (VAFF) system to provide forecasts when ash falls are expected to affect large areas. However, the present system predicts only the areas where ash may fall. A survey of VAFF users conducted in 2012 by JMA in areas close to active volcanoes (e.g. Asamayama, Kirishimayama, and Sakurajima) demonstrated strong demand for forecasts of the quantity of ash falls, prediction of areas of expected lapilli falls, and early issue of VAFFs. Recent studies by the Meteorological Research Institute (MRI) of JMA and an upgrade of the JMA supercomputer system will allow JMA to improve the current system by adding predictions of the quantity of ash falls and information about expected lapilli falls.

Volcanic ash falls affect buildings and many areas of human activity, such as traffic, infrastructure lifelines, agriculture, forestry, fisheries, and human health (e.g. Mt. Fuji Hazard Map Examination Committee [2004]; Horwell and Baxter [2006]; Wilson et al. [2012]). The hazards differ among these areas, and also according to the quantity of ash falls. Prediction of the quantity of ash falls is important for disaster mitigation or prevention. In March 2013, JMA presented a Proposal for Advanced VAFF at a meeting of the Working Group on Advanced VAFF. Based on that proposal, JMA will update the existing VAFF system to incorporate new capabilities and create a distribution system for forecasts. JMA plans to have the new system operational in spring 2015.

PRESENT VAFF SYSTEM

Under the present system, VAFFs are issued by JMA 30 to 40 minutes after large eruptions. The criteria for issuance of VAFFs are provided in Table 1. The VAFF is a graphical format including forecasts of the areal extent of ash fall at one-hour intervals for six hours after the eruption (Figure 1). VAFFs are sent to authorities at affected prefectures, disaster prevention agencies, and JMA branch offices and posted on the JMA website.

Table 1: JMA Criteria for issue of VAFFs (effective March 2014)

Volcano	Criteria
Asamayama	Observation of an eruption of plume height ≥3000 m above the crater rim. For plume heights of 1000–3000 m, a VAFF may be issued under some circumstances.
Kirishimayama (Shinmoedake)	Observation of an eruption of plume height ≥2000 m above the crater rim.
Sakurajima	Observation of an eruption of plume height ≥2500 m (criterion was 3000 m until May 20, 2013) above the crater rim.
Other Japanese volcanoes	Observation of an eruption plume height of ≥3000 m above the crater rim.

Hasegawa *et al.*

Hasegawa *et al.* *Journal of Applied Volcanology* 2015 **4**:2, doi:10.1186/s13617-014-0018-2

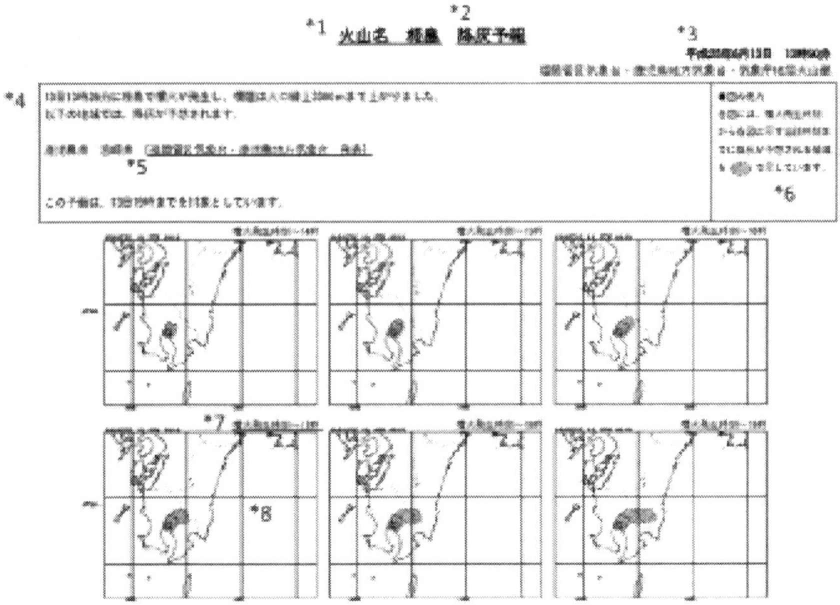

Figure 1: Example of a VAFF as issued under the present system. Translations from Japanese as numbered below. *1 Volcano name: *Sakurajima volcano.* *2 Forecast title. *3 Time of issue: *13:50 JST June 13, 2013.* *4 Summary: *An eruption at 13:26 (JST) on June 13 produced a plume that extended to 3300 m above the crater. Ash may fall in Kagoshima and Miyazaki Prefectures. This forecast is valid until 19:00 (JST) on June 13.* *5 Responsible observatory: *Fukuoka Regional Headquarters and Kagoshima Local Meteorological Office* *6 Legend. "red filled area": Forecast areal extent of ash fall at approximately 1-h intervals after onset of eruption. *7 Time at which map of forecast extent of ash fall (*8) applies: *17:00 (JST) on June 13.* *8 Forecast areal extent of ash fall.

VAFFs have been issued for eruptions at Asamayama, Kirishimayama, and Sakurajima volcanoes from March 2008 to March 2014 (Figure 2 and Table 2).

Figure 2: Volcanoes for which JMA issued VAFFs between March 2008 and March 2014. *Shinmoedake crater of Kirishimayama volcano.

Table 2: Number of VAFFs issued by JMA from March 31, 2008 to March 31, 2014

	Asamayama	**Kirishimayama (Shinmoedake)**	**Sakurajima**
2008 (Mar 31–Dec)			2
2009	1[1]		3
2010			
2011		39[2]	1
2012			4
2013			52[3]
2014 (Jan–Mar)			6

[1]Eruption on February 2, 2009.

[2]Eruptions from January 26 to April 18, 2011.

[3]Criterion for issue of VAFF changed on May 20, 2013.

Hasegawa *et al.* *Journal of Applied Volcanology* 2015 4:2, doi:10.1186/s13617-014-0018-2

CONTRIBUTIONS TO IMPROVEMENT OF THE VAFF SYSTEM

Improvements to Ash-fall Forecasts

Predictions of the areas and quantities of ash fall, and of areas of lapilli fall, have been improved in accuracy as a result of recent studies by MRI (Hashimoto et al. [2012]; Shimbori et al. [2014]). These studies document the use of weather radar systems to accurately measure the heights of volcanic plumes, which provide important initial data for predictive use. The recent upgrade of the JMA supercomputer system has also improved the precision of ash-fall and lapilli-fall predictions.

MRI applied the methodology described by Shimbori et al. ([2014]) to the 2011 eruption from the Shinmoedake crater of Kirishimayama volcano to derive predicted ash-fall data that matched observed data reasonably well (Figure 3; Shimbori et al. [2014]). JMA will incorporate this methodology in the new VAFF system.

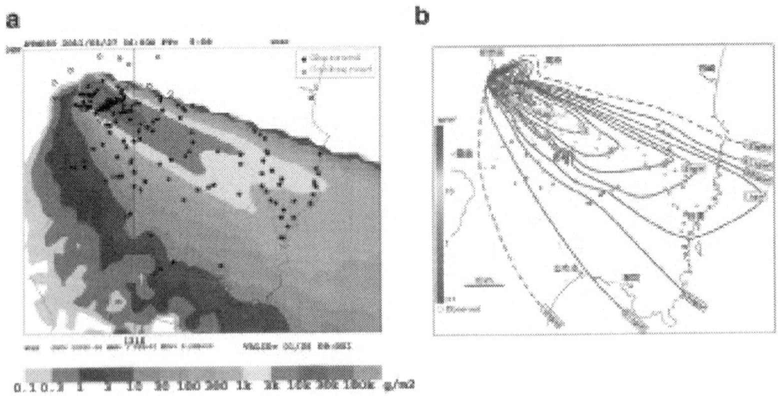

Figure 3: Comparison of ash-fall quantity forecast using MRI method and observed ash-fall quantity. (a) Quantitative tephra-fall predictions (g/m²) from 15:00 JST on January 26 until 00:00 JST on January 28 based on JMA-RATM

for the 2011 eruption at Shinmoedake volcano. Initial values for plume height and eruption duration based on data from the JMA weather radar network (Shimbori et al. [2014]). Legend: •, ash fall observed; ○, no ash fall observed. (b) Ash-fall observations (kg/m^2) for eruptions at Shinmoedake volcano during January 26–27, 2011 (Furukawa [2011]). Colored dots correspond to the color bar at the left side in the figure (b).

Technical methods of the VAFF are as follows: We first estimate the total mass of tephra from an eruption on the basis of plume height and eruption duration (Sect. A.2 of Shimbori et al. ([2010]) based on Suzuki ([1990])) and then use the JMA Regional Atmospheric Transport Model (JMA-RATM) driven by the grid point values of the numerical weather prediction model to predict the quantity of ash fall (thickness of predicted ash layer), the extent of the area of ash fall and the area in which lapilli are expected to fall (Figure 4).

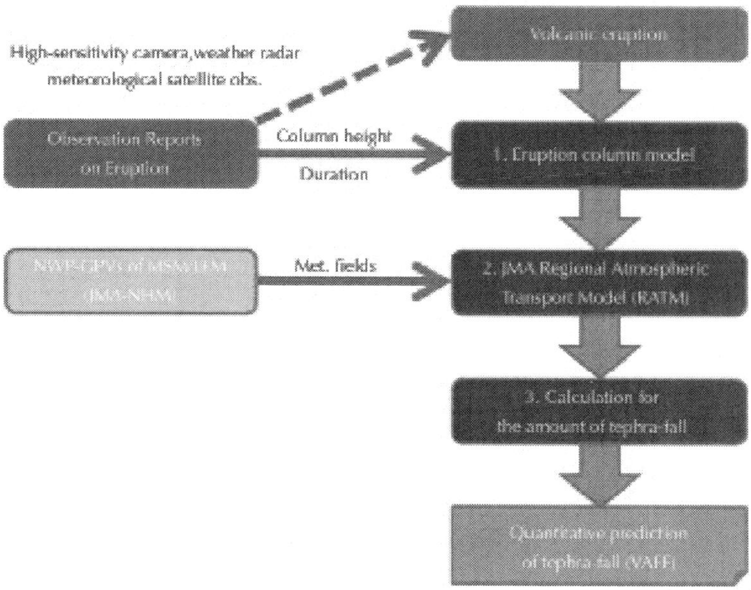

Figure 4: Flowchart of procedure for tephra-fall predictions using the JMA Regional Atmospheric Transport Model. Meso-scale Model (MSM) and Local Forecast Model (LFM) are gridded weather data (e.g., wind speed and direction, air temperature, atmospheric pressure) calculated by using the JMA numerical weather prediction model based on the Non-hydrostatic Model (NHM).

Survey to Determine Needs of VAFF Users

During March to September 2012, JMA carried out a survey to determine the needs of VAFF users in areas near volcanoes where there have been recent ash falls. Participants included disaster prevention organizations near Sakurajima, Kirishimayama, and Asamayama volcanoes (227 responses), local communities near those three volcanoes (386 responses), and local government authorities near 47 other active volcanoes in Japan (116 responses). The 47 volcanoes were selected on the recommendation of the Coordinating Committee for Prediction of Volcanic Eruption, a consultant body that advises the Director General of JMA.

Questions in the survey addressed VAFF users' needs in terms of the specific types of information they require, the timing of issue of VAFFs, and their means of access to VAFFs. Respondents were asked to consider their needs for VAFFs issued at three stages: before an eruption, immediately after the onset of an eruption, and after an eruption.

The result of the survey showed strong demand for early forecasts of the onset time and severity of ash fall and the areas to be affected, and information about areas to be affected by lapilli fall. The full survey results were published by JMA ([2013]).

Working Group on Advanced VAFF

JMA formed a Working Group on Advanced VAFF, which held meetings on July 5 and November 8, 2012, and February 19, 2013. The 16-member group included representatives from disaster prevention organizations, agricultural organizations, medical and broadcasting services, local government bodies, and volcanologists. The working group considered the results of the 2012 survey and sought to identify ways of improving the VAFF system to provide more-effective disaster mitigation. Matters discussed included how and what information should be provided, the conditions under which VAFFs should be issued, and appropriate forecasts for affected communities for various quantities of ash fall.

Outcomes of the meetings were published in March 2013 (Working Group on Advanced VAFF[2013b]) and are summarized below.

VAFFs should be issued when the forecast quantity of an ash fall is likely to affect local communities or the utilities they depend on in their daily life.

VAFFs should be easy for people to understand and act upon, and should include information about forecast lapilli falls.

VAFFs should be issued at three stages:

Before a possible eruption: regular information, including forecasts of areas that may be affected by ash fall and the likelihood of lapilli fall.

As soon as possible after the onset of an eruption: preliminary release of a brief forecast of the quantity of ash fall, areas to be affected, and the likelihood of lapilli fall.

After an eruption: when sufficient data are available, a detailed forecast of when ash fall will start and its quantity and areas to be affected.

VAFFs should categorize ash falls according to quantity, describe the potential impacts of each category on human health, transport systems, and utilities, and provide advice as to the appropriate actions to be taken by the general public during and after ash falls.

There should be a trial VAFF distribution to local government bodies near Sakurajima volcano incorporating the recommended improvements to the VAFF system and seeking their feedback.

Case studies are needed that illustrate the effects and appropriate actions for the general public for different quantities of ash fall.

Where applicable, results of volcanological studies undertaken at universities and other research institutions should be used to improve the VAFF system.

FUTURE VOLCANIC ASH FALL FORECASTS

JMA is preparing to implement the improved VAFF system based on the Proposal for Advanced VAFF (Working Group on Advanced VAFF [2013b]). The major improvements (Table 3) are as follows:

Table 3: Comparison of present and proposed VAFFs

	Present VAFF	Proposed VAFF		
Type and timing of forecast	Volcanic Ash Fall Forecast	Regular Information	Preliminary Forecast	Full Forecast
	Issued within 30–40 min of onset of eruption	Issued regularly (e.g. 3 hourly) before possible eruption	Issued within 5–10 min of onset of eruption	Issued within 20–30 min of onset of eruption
Contents	Forecast[1] of situation at six 1-h intervals after onset of eruption: Ash fall area. Prefectures where ash may fall	Forecast[1] for an assumed eruption: Ash-fall area. Lapilli-fall area. Recommended actions for affected communities	Forecast[2] for first hour after onset of eruption: Ash-fall area and quantity. Municipalities affected. Lapilli-fall area. Recommended actions for affected communities	Forecast[2] of situation at six 1-h intervals after onset of eruption: Ash-fall area and quantity. Municipalities affected. Start of ash fall for each municipality. Recommended action for affected communities

[1]Meso-scale Model used for forecasts (see Figure 4).

[2]Local Forecast Model used for forecasts (see Figure 4).

Hasegawa *et al.*

Hasegawa *et al. Journal of Applied Volcanology* 2015 **4**:2, doi:10.1186/s13617-014-0018-2

Sequential release of three types of VAFF: regular information issued before an possible eruption, a brief preliminary forecast issued 5–10 minutes after an eruption starts, and a full forecast issued 20–30 minutes after the eruption.

VAFFs will be issued in Japanese in both graphical and text formats.

JMA will provide a table categorizing ash falls by quantity (little, moderate, and much), documenting the likely effects on the general public, and advising them of necessary preparedness measures (Figure 5).

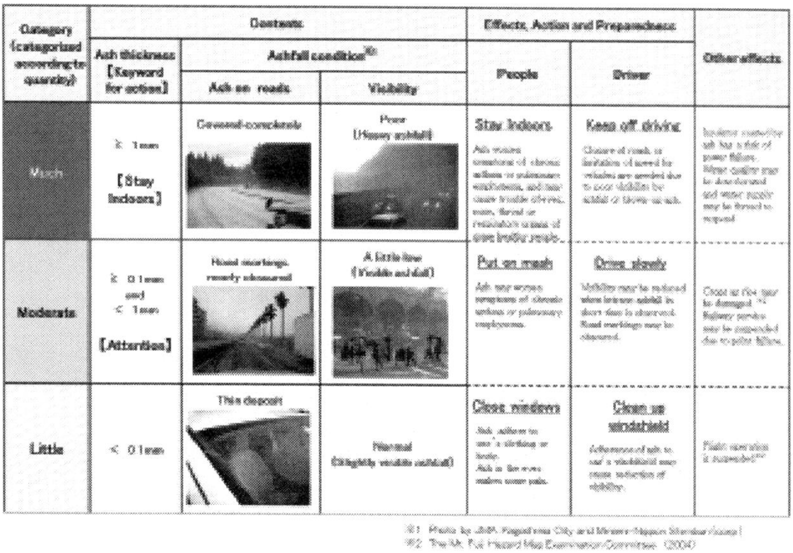

Figure 5: Categories of ash-fall quantity and corresponding conditions and preparedness measures. Table extracted from Working Group on Advanced VAFF ([2013b]) and translated into English.

VAFFs will be issued for ash-fall quantities in the moderate and much categories (Figure 5).

VAFFs will provide predictions of areas of lapilli fall (lapilli size ≥1 cm) separated from predictions of ash fall and provide recommended actions for people in areas affected, because ash fall and lapilli fall pose different hazards. Ash fall can, for example, cause a loss of traction on roadways whereas lapilli fall can break windshields of cars. Furthermore, some lapilli fall damage can occur even in areas of moderate ash fall. In addition, the prediction of area of lapilli fall was also strongly needed from users of VAFF (JMA [2013]).

Types of Forecast

Because the effects of ash fall manifest quickly after an eruption, VAFFs must be issued as early as possible. However, the information available at the start of an eruption lacks accuracy. Considering the needs of VAFF users and the technical limitations of ash-fall predictions, JMA will issue three types of VAFF, as described below (Figure 6).

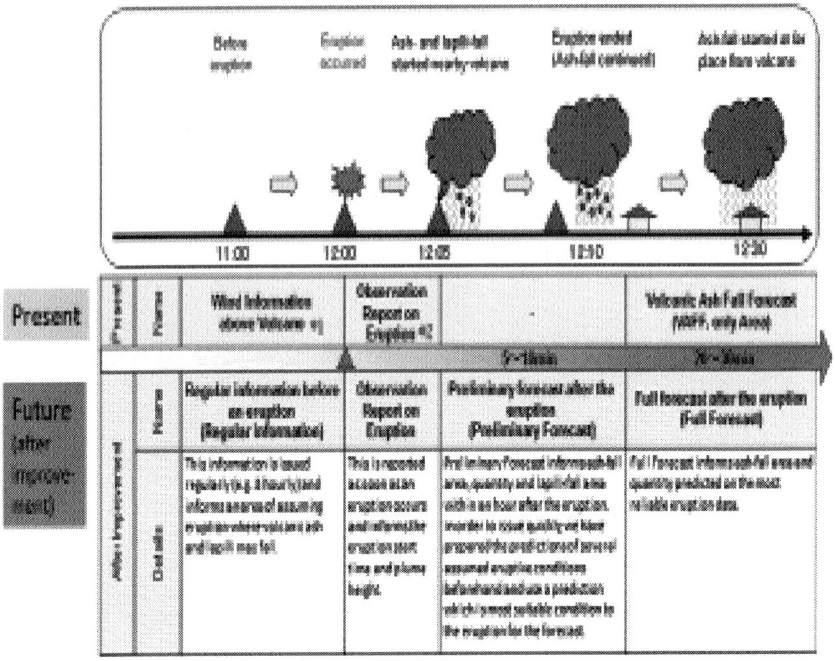

Figure 6: Flow diagram showing timing of issue and contents of the current VAFF and the proposed new VAFF. *1 JMA provides wind forecasts above volcanoes that have high likelihood of eruption. *2 JMA provides data such as eruption time, plume height, and direction of plume extension about 5 minutes after onset of an eruption.

Information Issued Before an Eruption (Regular Information)

This information will be issued regularly (e.g. 3 hourly) when there is a possibility that an eruption may occur to provide information on potential areas of ash fall and lapilli fall so that local communities can prepare (Figure 7).

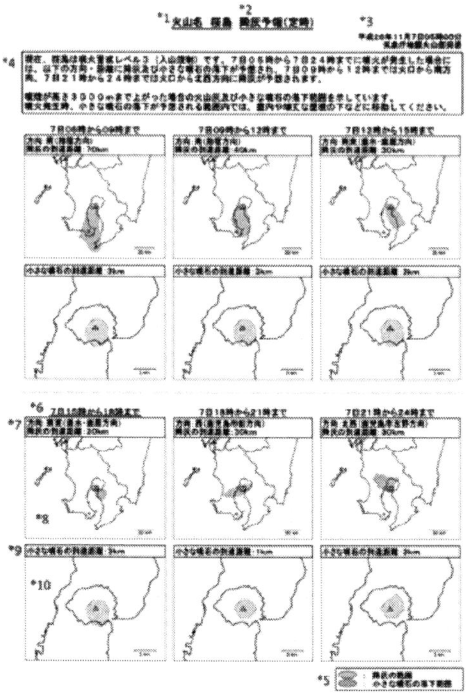

Figure 7: Sample of Regular Information (graphical format) in the new VAFF format. Translations from Japanese as numbered below. *1 Volcano name: *Sakurajima volcano.* *2 Information title. *3 Time of issue and source: *Issued at 05:00 JST on November 7, 2014, by Seismology and Volcanology Department of JMA.* *4 Summary: *Forecasts of areal extents of ash fall and lapilli fall for possible eruptions at three-hour intervals between 06:00 and 24:00 JST on November 7 are provided by the maps below. The forecast of areal extent of ash fall between 09:00 and 12:00 JST is forecasted to southern direction and that between 21:00 and 2400 JST is forecasted to north-west direction. Forecast areas of ash fall and lapilli fall are based on an eruption producing a 3000 m vertical plume above the crater. People in areas where lapilli fall is forecast should stay indoors or take refuge under a solid roof.* *5 Legend: Forecast areal extents of ash fall and lapilli fall. "Gray filled area": Ash fall. "Red filled area": Lapilli fall (≥1 cm size). *6 Annotation of time of eruption: *15:00 – 18:00 JST on November 7.* *7 Direction and distance from the crater to maximum extent of ash fall area. Name of localities affected is provided in parentheses: *Direction and distance are south-east and 20 km, respectively (direction of Tarumizu city and Kanoya city).* *8 Map showing forecast areal extent of ash fall. *9 Distance from the crater to maximum extent of area of lapilli fall: *3 km.* *10 Map showing forecast areal extent of lapilli fall.

Since the 1980s, JMA has issued current and forecast information about winds above Sakurajima volcano at regular intervals (e.g. twice daily) to allow local residents to prepare for possible ash fall.

Brief Forecast Issued Immediately After an Eruption (Preliminary Forecast)

The Preliminary Forecast will be issued within 5–10 minutes of the onset of an eruption. It will be based on JMA's standard 'Observation Report on Eruption' (eruption time, plume height and direction of movement) and will provide initial forecasts of ash-fall and lapilli-fall areas (Figure 8). The Preliminary Forecast will be important for communities close to the volcano.

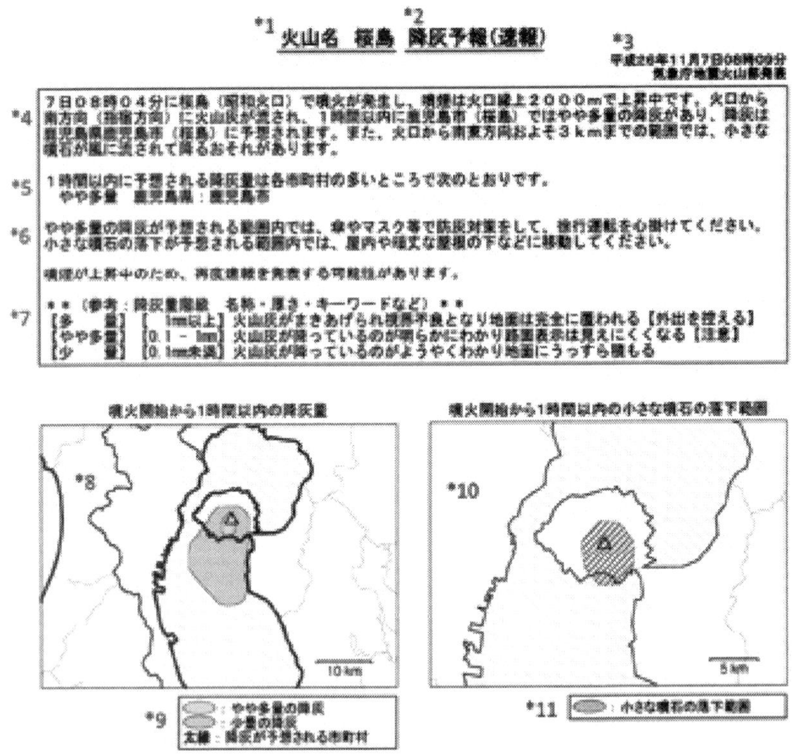

Figure 8: Sample of Preliminary Forecast (graphical format) in the new VAFF format. Translations from Japanese as numbered below. *1 Volcano name:

Sakurajima volcano. *2 Forecast title. *3 Time of issue and responsible observatory. *4 Summary: *An eruption occurred at 08:04 on November 7. A 2000 m eruption plume formed and is rising and extending to south. An ash fall of moderate quantity and areal extent of ash fall is forecast for Kagoshima city (Sakurajima) and areal extent of lapilli fall is forecasted to 3 km away from the crater to south-east direction is forecasted within one hour after onset of the eruption.* *5 Forecast of municipalities affected: *An ash fall of moderate quantity is forecast in Kagoshima city.* *6 Precautions for disaster prevention: *People in the open air in areas where a moderate ash fall is forecast should use an umbrella or wear a face mask to minimize their exposure to ash. People driving vehicles should drive slowly. People in areas where a lapilli fall is forecast should stay indoors or take refuge under a solid roof.* *7 Reference: Categories of ash fall quantity. Much: Ash deposits more than 1 mm thick; low visibility by ash fall and roads completely covered by ash; Stay Indoors. Moderate: Ash deposits 0.1 to 1 mm thick; recognizable ash fall and road markings partly obscured by ash; Attention. Little: thin ash deposits, less than 0.1 mm thick. *8 Forecast areal extent of ash fall. *9 Legend (ash fall). "Yellow filled area": Area of ash fall of moderate quantity. "Gray filled area": Area of ash fall of little quantity. Thick black line: boundary of municipality to be affected by ash fall. *10: Forecast areal extent of lapilli fall. *11: Legend (lapilli fall). "Red filled area": Area of lapilli fall (≥1 cm size).

To facilitate early issue, JMA has prepared in advance various model forecasts for different conditions (eruption time, plume height, wind direction). When an eruption starts, JMA will use the model forecast that best fits the observed conditions for the Preliminary Forecast.

Full Forecast Issued After the Eruption (Full Forecast)

The Full Forecast will be issued within 20–30 minutes of the onset of an eruption and will contain accurate forecasts of the area and quantity of ash fall. It will give communities living farther from the volcano time to make appropriate preparations for ash fall (Figure 9). For the Full Forecast, JMA will improve the accuracy of plume height estimates of the previously issued Preliminary Forecast by using weather radar and meteorological satellite data in addition to visual observations.

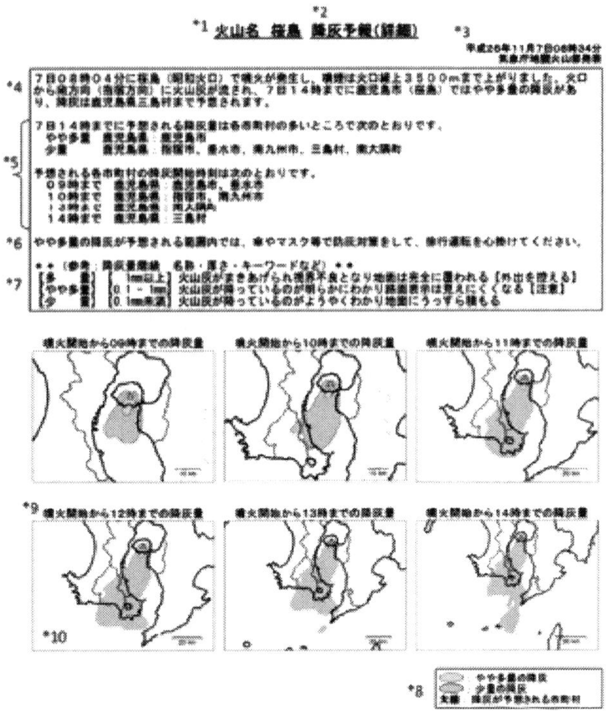

Figure 9: Sample of Full Forecast (graphical format) in the new VAFF format**.** Translations from Japanese as numbered below. *1 Volcano name: *Sakurajima volcano.* *2 Forecast title. *3 Time of issue and responsible observatory. *4 Summary: *An eruption occurred at 08:04 on November 7. A 3500 m eruption plume extended to south was observed. An ash fall of moderate quantity is forecast for Kagoshima city (Sakurajima), and areal extent of ash fall is forecasted to Mishima village.**5 Forecast of municipalities affected and start time and quantity of ash fall: *Forecasts of ash fall quantity and municipalities affected are as follows. Moderate: Kagoshima city. Little: Ibusuki city et al. Forecast starts of ash fall at municipalities are as follows. Before 09:00 on November 7: Kagoshima city and Tarumizu city. 09:00 to 10:00: Ibusuki city and Minami-kyusyu city. 12:00 to 13:00: Minami-osumi town. 13:00 to 14:00: Mishima village.* *6 Precautions for disaster prevention. *People in the open air in areas where moderate ash fall is forecast should use an umbrella or wear a face mask to minimize exposure to ash. People driving vehicles should drive slowly.* *7 Reference: Categories of ash-fall quantity. Much: Ash deposits more than 1 mm thick; low visibility by ash fall and roads completely covered by ash; Stay Indoors. Moderate: Ash deposits 0.1 to 1 mm thick; recognizable ash fall and road markings partly obscured by ash; Attention.

Little: thin ash deposits, less than 0.1 mm thick. *8 Legend: "Yellow filled area": Area of ash fall of moderate quantity. "Gray filled area": Area of ash fall of little quantity. Thick black line: boundary of municipality to be affected by ash fall. *9 Time of forecast shown below (*10). *10 Forecast areal extent of ash fall for each quantity category.

Format of VAFFs

VAFF information will be provided in the following formats.

Textual information will be provided in Extensible Markup Language (XML).

Graphical information will be provided in Portable Document Format (PDF) format.

For specialist users, quantitative forecasts (g/m^2) of ash fall will be provided as numerical data at the JMA website.

Categories of Ash-fall Quantity

Quantitative predictions of the amount of ash fall (expressed in g/m^2) include some errors and may be difficult for non-specialist VAFF users to understand. Therefore, ash-fall quantities will be expressed in three qualitative categories (little, moderate, and much), as proposed by the Working Group on Advanced VAFF. These categories will each be accompanied by estimates of the expected thickness of ash deposits, a simple description (including photos) of the conditions that communities may encounter after an ash fall (road conditions, reduced visibility), and recommended safety precautions to be taken by the general public (Figure 5).

Information about Lapilli Fall

The damage caused by lapilli of ≥1 cm size can be considerable (Figure 10 and Table 4). Because lapilli fall faster than ash particles, their arrival at the ground surface precedes that of ash particles. Even for small eruptions with small plumes, considerable damage can be caused by only a few large lapilli. JMA will therefore provide forecasts of areas of lapilli fall where lapilli of ≥1 cm size are expected.

Figure 10: Damage caused by lapilli fall. Top: Rear car window destroyed by lapilli fall (4 to 6 cm size) 7 km from the Shinmoedake crater of Kirishimayama volcano. Bottom: Individual lapillus. Individual squares on the ruler are 5 mm by 5 mm. Photos by JMA (January 27, 2011).

Table 4: Examples of damage due to lapilli fall after recent eruptions in Japan (since 1970)

Volcano	Eruption date (yyyy/mm/dd)	Size of lapilli	Damage
Sakurajima	1983/8/14	6–7 cm	Windshield of car, glass pane in front door of house
Sakurajima	1978/7/31	5–6 cm	Injuries to people, car window, house window
Sakurajima	1986/6/24	5–6 cm	Windshield of car, slate roof
Sakurajima	1979/1/5	4–6 cm	Windshield of car
Kirishimayama (Shinmoedake)	2011/1/27	4–6 cm	Window of car
Sakurajima	1983/5/26	4 cm	Windshield of car, slate roof
Sakurajima	2000/10/7	3 cm	Window of car, greenhouse
Asamayama	1973/2/20	3 cm	Window of car
Sakurajima	1986/8/16	2–3 cm	Windshield of car
Kirishimayama (Shinmoedake)	2011/4/18	2 cm	Solar panel and fittings

Sakurajima	1983/8/14	1–3 cm	Window of car
Kirishimayama (Shinmoedake)	2011/2/14	1.5–2 cm	Slate roof

Table translated into English from Working Group on Advanced VAFF ([2013a]).

Hasegawa *et al.*

Hasegawa *et al. Journal of Applied Volcanology* 2015 4:2, doi: 10.1186/s13617-014-0018-2

Trial distribution of proposed new VAFF

Since April 2013, JMA has trialed distribution of the proposed new VAFF (as described above and based on real eruption data) to local government authorities in the region of Sakurajima volcano. In this region, the residents of Kagoshima city (population ~600,000) have experienced frequent ash falls for several decades. JMA will use feedback from the trial to further refine the contents of VAFFs and the conditions under which they are issued.

FURTHER REFINEMENTS OF VAFF SYSTEM

JMA and MRI will continue to work together to refine the VAFF system. Areas to be addressed include:

Improvement of RATM-based ash-fall predictions for strong wind conditions and for large-scale eruptions

Development of a method to estimate volcanic plume height, eruption column mass and eruption duration at times when bad weather prevents visual observations and observations using radar data and satellite imagery

Regular examination of the relationship between ash-fall quantity and its effect on people, and review of the recommended actions for affected communities.

Regular review of the three categories of ash-fall quantity and the thresholds that separate them (Figure 5)

DISCUSSION

The Working Group on Advanced VAFF ([2013b]) recommended further work on the accuracy of the JMA ash-fall prediction model RATM, for periods of strong winds and review of the methodology used to calculate the mass of ash in a volcanic plume. These two issues are discussed below.

Effect of Strong Winds on Forecast Accuracy

JMA reviewed the accuracy of lapilli-fall forecasts provided in trial VAFFs issued for eight eruptions of Sakurajima volcano between April 2013 and February 2014. Lapilli falls were observed more than 3 km from the craters in all cases. With one exception, the actual areas of lapilli fall agreed well with forecasts. The errant forecast was for an eruption on September 4, 2013 (Figure 11), where the forecast area of lapilli fall was clearly smaller than the observed area. During this eruption, wind speed about 500 m above the volcano was 13 m/s, whereas wind speeds during the other seven eruptions were 5–10 m/s.

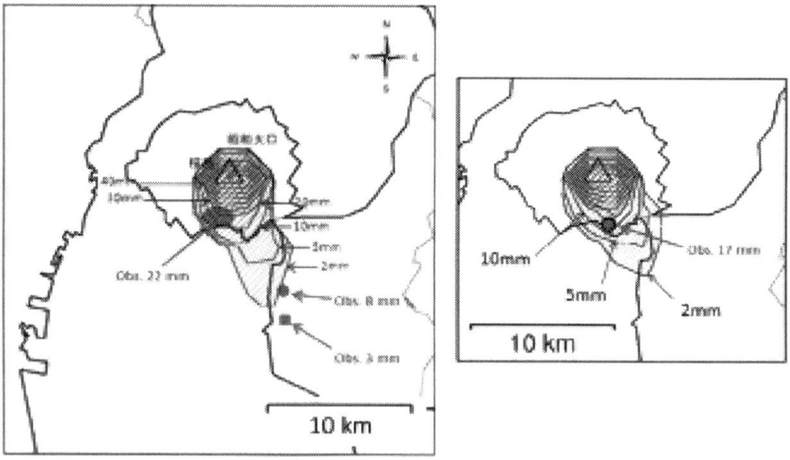

Figure 11: Comparisons of predicted and observed areas of lapilli fall for two eruptions at Sakurajima volcano. Left: Underestimation of area and lapilli clast size of lapilli fall for an eruption at 11:00 JST on September 4, 2013. The plume rose 2800 m above the crater and wind speed at about 500 m above

the crater at the time of the eruption was 12.6 m/s. Right: Predicted and observed areas and lapilli clast size of lapilli fall in agreement. Eruption at 14:37 JST on January 22, 2014. The plume rose 2800 m above the crater and wind speed at about 500 m above the crater at the time of the eruption was 6.3 m/s. Red: Points or areas where lapilli fall was observed. Obs: maximum size of observed lapilli. Yellow: Predicted area of lapilli fall and size of lapilli; contours and their annotations indicate size of lapilli.

For initial predictions of areas of ash fall, the JMA-RATM assumes that ash plumes form a vertical column directly above the crater (Shimbori et al. [2010]). Under the influence of strong winds, an ash plume (and the lapilli it contains) may drift leeward and diverge from the vertical, thus resulting in lapilli fall over a greater area than predicted, as was suggested by the Working Group on Advanced VAFF ([2013b]). Although we do not have sufficient data to allow us to examine ash falls under strong wind conditions, we consider that strong winds will reduce the accuracy of our ash-fall predictions.

The current model for plume formation needs revision to take into account the effect of strong winds (Suzuki and Koyaguchi [2013]). In addition, local topographic effects near the volcano, assumed grain-size distribution, and vertical distribution of lapilli in the initial eruption column model also need to be considered adequately. Evaluation of these factors should be the subject of future research.

Estimation of Mass of Volcanic Ash in ash Plumes

The JMA-RATM estimates the mass of volcanic ash in a plume by applying a formula based on empirical data from past eruptions in various parts of the world (plume height, eruption duration, and field data on actual areas of ash fall; Sect. A.2 of Shimbori et al. ([2010]) To date, we have applied this model regardless of plume size; however, the Working Group on Advanced VAFF ([2013b]) suggested that this approach might not be accurate for small eruptions.

Between April 2013 and February 2014 there were only four eruptions at Sakurajima volcano for which there were sufficient observations to allow us to compare observed and predicted masses of ash generated. Although our estimate was accurate for the largest of these eruptions (16:31 JST on August 18, 2013), which generated an ash plume that rose 5000 m above the crater, we underestimated the masses of ash for the three smaller eruptions, which had plume heights of 2400–3300 m.

These results suggest that the estimates of the masses of ash in plumes that we obtained by using the current approach are inaccurate for small eruptions. As suggested by the Working Group on Advanced VAFF ([2013b]), further study is needed to improve these estimates.

CONCLUSIONS

To improve the current VAFF system and make it a more-effective tool for disaster prevention or mitigation, JMA took into account the needs of VAFF users and the opinions of experts in various fields including disaster prevention organizations. The current VAFF provides only forecasts of areas of ash fall. Important improvements to be implemented in the new VAFF system are the provision of predictions of ash-fall quantity and areas of lapilli fall. Two factors that have made important contributions to these improvements are advances in the methodology for prediction of ash-fall quantity and areas of lapilli fall as a result of studies by MRI, and increased processing capacity of the JMA supercomputer system following a recent upgrade.

The main improvements incorporated in the new VAFF system are as follows.

Three types of VAFF will be sequentially issued: information issued before an eruption (Regular Information), a brief forecast issued within 10 minutes of the onset of an eruption (Preliminary Forecast) and a more extensive and accurate forecast issued within 30 minutes of the onset of the eruption (Full Forecast).

VAFFs will be issued in Japanese in both graphical format and text format.

JMA will provide with VAFFs a table that categorizes ash fall by quantity, documents the potential effects on people for each category and the preparedness measures they should take.

VAFFs will be issued for ash falls of moderate and much quantity.

In addition to forecasts of areas of ash fall provided in current VAFFs, the new VAFFs will include predictions of ash-fall quantity and areas of lapilli fall (for lapilli size ≥1 cm), and recommended precautions for people to take during and after ash and lapilli falls.

JMA plans to start using the new VAFF system in spring 2015, and will continue to review and make further improvements in the future.

AUTHORS' CONTRIBUTION

YHas guided the review process of the research, and is the main author of the manuscript. AS contributed to making of format of new VAFF and the categories of ash fall quantity. YoHay contributed to making of format of new VAFF and the categories of ash fall quantity. YuHay contributed to making of format of new VAFF and the categories of ash fall quantity. SS contributed to making of format of new VAFF and the categories of ash fall quantity. TS contributed to improvement of techniques of ash fall prediction. All authors read and approved the final manuscript.

ACKNOWLEDGEMENTS

We are grateful to the chairman (Prof. A. Tanaka) and members of the Working Group on Advanced VAFF for their contributions to this project. We thank the people who participated in our survey of the needs of VAFF users near Asamayama, Kirishimayama, and Sakurajima volcanoes, and the organizations that cooperated in our trial distribution of VAFFs near Sakurajima volcano, including local governments and local media. We also thank two anonymous reviewers. Advice about improvements of the VAFF system from Dr. H. Yamasato, S. Harada, and other JMA staff members is much appreciated.

REFERENCES

1. Furukawa R (2011) Eruption of Shinmoedake volcano, Kirishima volcano group, 2011. AIST Today 2011–09:15
2. Hashimoto A, Shimbori T, Fukui K (2012) Tephra fall simulation for the eruptions at Mt. Shinmoe-dake during 26–27 January 2011 with JMA NHM. SOLA 8:37-40 doi:10.2151/sola. 2012-010

3. Horwell CJ, Baxter PJ (2006) The respiratory health hazards of volcanic ash: a review for volcanic risk mitigation. Bull Volc 69:1-24
4. (2013) Report of Survey of Needs for Improvement of VAFF (in Japanese).
5. (2004) Report of Mt. Fuji Hazard Map Examination Committee (in Japanese).
6. Shimbori T, Aikawa Y, Fukui K, Hashimoto A, Seino N, Yamasato H (2010) Quantitative tephra fall prediction with the JMA mesoscale tracer transport model for volcanic ash: A case study of the eruption at Asama volcano in 2009 (in Japanese). Pap Met Geophys 61:13-29 doi:10.2467/mripapers.61.13
7. Shimbori T, Kai R, Hayashi Y, Hayashi Y, Sugai A, Hasegawa Y, Hashimoto A, Takagi A, Yamamoto T, Fukui K (2014) Tephra fall predictions with the JMA regional atmospheric transport model: A case study of the eruptions at Shinmoe-dake volcano in 2011 (in Japanese). Pap Met Geophys 65:75-107 doi:10.2467/mripapers.65.75
8. Suzuki T (1990) Potential hazards of pyroclastic air-fall (in Japanese). Bull Inst Vocational Training 19A:1-17
9. Suzuki YJ, Koyaguchi T (2013) 3D numerical simulation of volcanic eruption clouds during the 2011 Shinmoe-dake eruptions. Earth Planets Space 65:581-589 doi:10.5047/eps.2013.03.009
10. Wilson TM, Stewart C, Sword-Daniels V, Leonard GS, Johnston DM, Cole JW, Wardman J, Wilson G, Barnard ST (2012) Volcanic ash impacts on critical infrastructure. Phys Chem Earth 45–46:5-23 doi:10.1016/j.pce.2011.06.006
11. (2013) Materials of 3rd Meeting of the Working Group on Advanced VAFF (in Japanese).
12. (2013) Proposal for Advanced VAFF (in Japanese).

Chapter 3

Beyond the Volcanic Crisis: Co-governance of Risk in Montserrat

Emily Wilkinson

Overseas Development Institute, Climate and Environment Programme, 203 Blackfriars Road, London SE1 8NJ, UK

ABSTRACT

Disaster risk governance is concerned with how institutions change in response to perturbations or, conversely, are able to remain static for long periods of time. In Montserrat, the volcanic eruption in 1995 produced unprecedented challenges for both local government authorities and the UK Government. The sharp and sustained rise in the level of volcanic risk combined with an inadequate response from UK and local authorities prompted a shift in governance arrangements, and when levels of risk declined these new configurations did not go back to their pre-crisis state.

This paper focuses on one aspect of this governance transition: the relationship between the local Montserratian government and the UK government. Before the eruption Montserrat enjoyed high levels of political and fiscal independence from the UK in disaster risk management and other investment decisions, but the volcanic crisis highlighted low levels of capacity and the inherent instability in this system. A new co-governance regime was established after the crisis, characterised by greater UK intervention in local investment decisions and some loss of political sovereignty. On the other hand, Montserrat has become more integrated in regional and international disaster risk governance systems, and today the division of local and central responsibilities for different aspects of disaster risk management is much clearer than before the volcanic crisis.

This paper demonstrates how disasters can create spaces for existing risk governance systems to be questioned and modified. The volcanic crisis led to a reconsideration of responsibilities and risk management practices by both Montserratian and UK authorities, and initiated a process of transformation in land-use and development planning that has substantially reduced levels of volcanic risk on the island. However, these benefits have to be weighed against loss of livelihoods for a significant proportion of the population and considerable social upheaval.

Critical to the success of this new development model is the need for vertical coherence and dialogue between different stakeholders. Montserrat and UK risk governance systems are more integrated now, but unless communities are engaged in risk management decisions, Montserrat's low- risk development model could come unstuck. Small islands with large risks can learn a lot from the Montserrat experience.

BACKGROUND

A disaster risk governance system comprises a complex web of actors and networks involved in formulating and implementing polices to manage disaster risk, institutional arrangements that determine the relationships, roles and responsibilities of these actors, coordinating mechanisms and political culture, including different perceptions of risk

(Renn [2008]; Wilkinson [2013]). The system is therefore characterised by a number of elements of interaction such as stakeholder participation in policies to reduce risk (Pelling [2011]).

This paper is concerned with multiple levels of risk governance and (a) whether crisis leads to changes in the system; (b) the nature of the shift (key aspects of the system that are altered); and (c) the change process. This is assessed in the context of Montserrat in the British West Indies, where a long-duration volcanic crisis in the 1990s highlighted internal contradictions inherent in the broader system of governance. Other volcanic eruptions in the eastern Caribbean have resulted in greater loss of life than the 1995–1997 eruption of Soufrière Hills Volcano–most notably the 1902 eruption of Mont Pelée in Martinique, which killed 29,000 people (Tanguy [1994]) – but the Montserrat volcanic crisis has attracted special interest from natural and social scientists alike because of the unpredictable and incremental escalation of volcanic activity, coupled with vulnerability and exposure characteristics found only on small islands.

A series of forced evacuations and delineation of risk zones in Montserrat avoided the mass casualties of Martinique, but on 25 June 1997, 19 people returning to the exclusion zone were killed by pyroclastic flows (see Figure 1). Two months later pyroclastic flows engulfed the capital, Plymouth, putting an end to discussions on whether the port and other major facilities could be used for the foreseeable future. Before these tragic events the Government of Montserrat and the UK Government had been struggling to manage a crisis for which they were relatively unprepared and when rehabilitation and reconstruction began, they continued to face difficulties and public criticism.

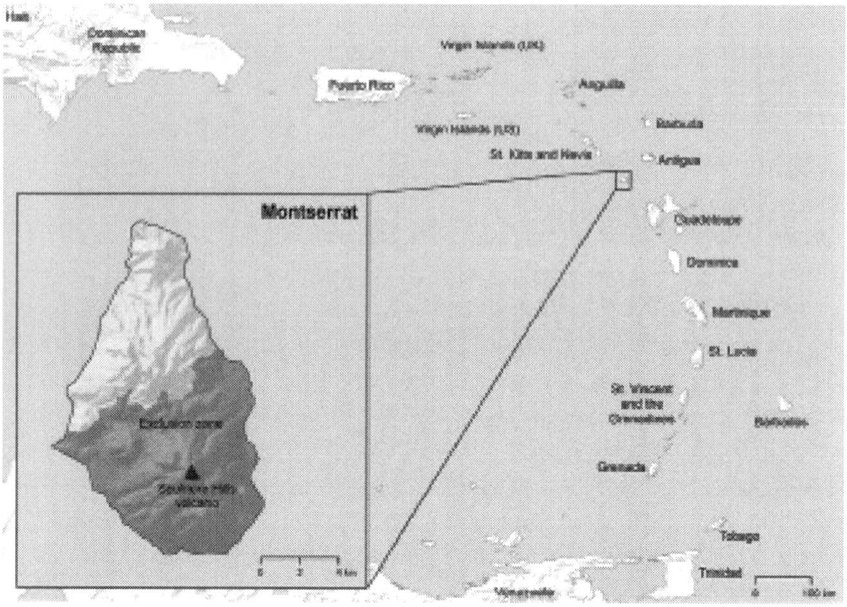

Figure 1: Map of Montserrat. Katy Mee, British Geological Survey.

The purpose of this paper is not to provide a summary of events, or attempt to evaluate the effectiveness of collective responses to the volcanic crisis. Numerous reports and papers have been commissioned and written on the events and decisions taken by UK and Montserrat authorities, each presenting a view of what went wrong (see, for example, Clay *et al.* [1999]; Donovan *et al.* [2012]; Kokelaar [2002]). More recently, studies have focussed on post-disaster reconstruction and the UK government's performance in promoting long-term development on the island (ICAI [2013]; Sword-Daniels *et al.* [2013]). Rather, the paper takes a longer term view of changes in disaster risk governance, analysing the critical shifts that occurred in response to an extreme event, as well as the change processes themselves.

Conceptual Framework

Concepts of 'systems' and 'scale' are used to study the nature of changes in response to perturbations. These draw on the socio-economic systems, resilience and natural resource governance literature. 'Scale' is defined as the spatial, temporal, quantitative, or analytical dimensions used to measure and study any phenomenon (Gibson *et al.*[2000]). This paper focusses on two scales: the jurisdictional scale, which can be divided into different bounded and organised political units, with linkages between them; and the temporal scale, which can be divided into different 'time frames' related to rates, durations, or frequencies (Ostrom *et al.* [1999]). Effective disaster risk management (DRM) depends on the cooperation of international, national, regional and local institutions across temporal and jurisdictional and geographical scales, so understanding these interactions is critical.

Within the jurisdictional scale, the vertical power relationships between local and central political units are of particular interest. Building on work by Claudia Pahl-Wostl ([2009]) on multi-level learning processes and adaptation, two key aspects of vertical governance dialectics can be identified: (a) dispersion of decision-making authority and (b) vertical coherence.

(a) The dispersion of decision-making authority refers to the authority of different centres of decision making that are formally independent of each other (Ostrom, [1997]; McGinnis [2000]; Heinelt [2002]). Local governments are thought to require autonomy from higher levels of government so that they can define their own priorities and implement DRM measures without too much interference, and thus gain credibility and trust from their citizens; both of which have proven critical for disaster risk management (Wilkinson [2012]). However, this kind of autonomous, inclusive style of governance is not the *modus operandi* of most local governments. This paper focusses on the political and fiscal autonomy of the Montserrat government to develop its own risk management policies, as well as its capacity to do so. It discusses the evolution of co-governance arrangements through which UK authorities and the Montserrat government jointly make decisions to manage exposure to volcanic risk.

(b) Vertical coherence is concerned with the division of roles and responsibilities for risk management between different political units, from local government to tiers higher up the scale – including provincial governments (or states in federal systems of governance), national government and regional authorities like the European Union (EU). Incoherence in service delivery often occurs because of poorly defined and overlapping mandates (resulting in omission and/ or replication in service delivery), overly complex structures (Pahl-Wostl [2009]), capacity constraints and unfunded mandates (Posner [1998]), as well as centrist and paternalistic tendencies in public administration systems (Wilkinson [2012]); all of which can constrain progress on managing disaster risk. Hence, an alignment of interests between governance scales can help to promote more effective DRM. This includes not only the participation of actors from one level in decision-making processes at another but also institutions and knowledge produced at one level influencing processes at another (Pahl-Wostl [2009]).

Particularly critical for DRM is the alignment of risk perceptions across scales of governance through bi-directional process (Slovic [1987]). Different people and cultures respond to disaster risk differently (Gaillard [2008]; Heijmans [2001]; Paton et al. [2001]; Paton *et al.*[2010]), and in the context of volcanic hazards, proximity to the hazard (Gregg *et al.* [2004]), living memory of an eruption and level of impact (Paton *et al.* [2001]) all shape risk perceptions amongst individuals and groups. Even if risk perception is high, people may still put concerns about convenience and living costs ahead of their desire to lessen their exposure (Gaillard [2008]). This suggests that the idea of an *acceptable* level of risk is inappropriate. Rather, people do not accept risks but tolerate them to secure certain benefits (Pidgeon *et al.* [1992]; Simmons and Walker [1999]). The values underlying any notion of tolerable risk may not be shared by everyone; in fact much research on risk analysis and societal reactions to different threats highlights the differences between institutional responses such as regulation and public responses (Barnes [2002]). In particular, disaster events can result in the creation of new official rules to control risk that minimise exposure - for example through the creation of exclusion zones and resettlement policies - but these values may not be shared by those that live in exposed areas and whom are to be resettled. Dialogue and negotiation between authorities and communities is therefore required to reach more sustainable solutions (Haynes *et al.*[2008]).

The analysis of the disaster risk governance system in this paper draws heavily on resilience thinking, and in particular resilience frameworks that emphasise the capacity of a system to respond to shocks and stresses in different ways – such as by coping, adapting and transforming (Bené *et al.* [2012]; Cutter *et al.*[2008]; Pelling [2011]). Disaster risk governance systems are highly sensitive to rates, durations and frequencies of disaster events and changes in the system often occur as a result of these events as well as in response to other external pressures. The feedback processes are however non-linear and unpredictable (Ramalingam *et al.* [2008]).

Notwithstanding their idiosyncrasies, volcanic eruptions can be characterised in terms of their spatial and temporal dimensions: they are often slow onset and long duration event that allow for changes in policy and behaviour while the event is still unfolding. It is usually possible to identify sharp increases in the level of risk, resulting in crisis period(s) for affected populations and decision-makers. Hence volcanic disaster risk can be considered to have three temporal phases within which feedback processes occur, with accompanying options for institutional learning and collective action:

- Pre-crisis period, in which action may be taken to mitigate existing and anticipate future risk, such as land-use planning, retrofitting roofs, the development and enforcement of building codes, education and training programmes. Land-use planning is a prospective tool that can be used to prevent or limit construction in unsafe areas, while relocation and re-zoning of space is a corrective tool to reduce existing exposure to hazards. Education related to building practices that reduce ash entry into homes is a risk management activity that anticipates and reduces risk in the future, while training on early warning systems manages current levels of risk by encouraging evacuations and reducing loss of life.
- Crisis period, which we can sub-divide into: a) start of the eruption and potentially long period of unrest (often characterised by seismic activity), which can be treated as a preparedness phase; and b) heightening of the crisis, usually initiated by an eruption, prompting emergency response activities to reduce negative impacts on people, such as food aid and shelter provision. These sub-phases vary widely across settings however and some volcanoes may do (a) and not (b), while some have (b) with no (a).
- Post-crisis period, characterised by short- and longer-term recovery measures (the first of which may commence during the crisis period) to restore livelihoods and infrastructure as well as control future risk and promote sustainability (Alexander [2002]; Tierney [2012]). These corrective and prospective risk reduction measures are more likely to occur in the post-crisis period than before an event has occurred as disasters highlight previous

failures and can act as catalysts for policy reform (Birkland [2006]).

These three temporal phases may overlap if time between subsequent eruptions is short. Also, the shift from one state to another is not necessarily demarcated by the volcanic hazards themselves: there may still be low-impact hazards occurring in in the post-crisis period; and changes in the level of risk might also be caused by non-volcanic events that alter the level of exposure or vulnerability to different hazards. Nor do the phases identified above represent a cycle in the social system (from stability-to crisis-returning to a stable state). Indeed, the concept of a 'disaster cycle' has been heavily criticised by social scientists for representing disasters as temporary interruptions of a linear development process and governance systems, after which society returns to normal (Christoplos *et al.*[2001]; Hewitt [1983]; Twigg [2004]). Governance systems do sometimes return to pre-crisis states, demonstrating the stability or persistence of institutions in the face of extreme social events (Schreyögg and Sydow [2010]). However, more often in environmental and social systems, regime changes occur following significant perturbations, whereby the system moves to another stable state and sometimes this regime shift is irreversible (Whitten *et al.* [2012]). Similarly, for disaster risk governance systems, we can expect disasters to alter components of the system, at least temporarily – whether perturbations are low-intensity but frequent or singular, high-intensity events. Changes in the disaster risk governance system during and following a protracted crisis can therefore be characterised in terms of their *stability*, from temporary alterations to permanent, irreversible shifts.

Another aspect of the change process is the *extent* to which the governance system is altered by the event – whether it undergoes fundamental changes or not. Levels of organisational change are described in the literature on adaptation and resilience, where differences are drawn between single and double-loop (and sometimes even triple-loop) learning; incremental and radical reform; transitions and transformations (Pelling [2011]). According to Mark Pelling ([2011]: 74) transitions or incremental changes can be seen when 'the aims and practices of geographically or sectorally-bound activities push but do not overturn established political regimes', while transformation 'is an extreme case where profound change alters the distribution of rights and responsibilities and visions of development across society'.

Similarly, while single-loop learning describes the detection of an error and correction without questioning the underlying values of the system, double-loop learning occurs 'when mismatches are corrected by first examining and altering the governing variables and then the actions' (Argyris [1999]: 68).

The nature of disaster risk governance shifts and change processes described above can be summarised in a matrix (see Table 1) and form the conceptual basis for analysing institutional change in Montserrat. The unique co-governance characteristics in Montserrat and other UK overseas territories present a number of challenges to studying institutional change of any sort. Institutions and individuals interact in ways that are very different to other governance settings[a], making generalisations or lesson drawing about drivers of change particularly difficult. However, in focussing on the dynamics of vertical governance, direct comparisons can be made to governance arrangements in other contexts, including in federal governance systems such as Mexico and India, decentralised systems such as those found elsewhere in the Caribbean and other multi-layered systems of governance, such as the European Union. Conclusions are tentative and caution must be applied in making generalisations, but the Montserrat case is instructive of a more permanent co-governance transition that can occur following a volcanic crisis.

[a]For a more detailed discussion of complex social relations and personal politics of small societies as well as the dependency mentality of overseas territories ad former colonies see Skinner ([2002]).

Table 1: Nature and processes of disaster risk governance change

Change in disaster risk governance system		Phases of volcanic risk		
		(i) Pre-crisis	(ii) During crisis	(iii) Post-crisis
Nature of change	(a) Dispersion of decision-making			
	(b) Vertical coherence			
Change process	(a) Stability			
	(b) Extent			

Wilkinson
Wilkinson*Journal of Applied Volcanology* 2015 4:3,doi:10.1186/s13617-014-0021-7

METHODS

The analysis of vertical governance arrangements in Montserrat presented below is based on qualitative primary data collected through a 'forensic' workshop[b] held in September 2012 with 70 participants representing five stakeholder groups: scientists, UK government officials, Montserrat government officials (including disaster managers), regional agency staff and community representatives. The aim was to explore components of resilience during and after the volcanic crisis as well as internal and external factors that have undermined it. Moderated focus group discussions on key events, tipping points and phases of change were held and recorded. In addition, 16 semi-structured interviews were conducted with local and UK government officials and community leaders. Workshop and interview recordings were transcribed and coded and analysed using Atlas-ti software.

[b]The Montserrat workshop was run by the STREVA programme as part of a 'forensic' research process, from 25-29th September 2012.

The coding categories were derived from the conceptual framework to capture data on: (i) risk management policies and key decisions taken during different time periods (before, during and after the crisis); (ii) roles and responsibilities of different actors for DRM activities; (iv) relationships between UK and local government authorities; and (v) public perceptions of government decisions on risk management policies (both UK and local). Interview and focus group data was also coded for issues of (vi) risk perception, (vii) trust and (viii) participation. Data was triangulated across the five stakeholder groups and with secondary literature, to help explain differences in judgements about decisions taken by local and UK authorities. Tensions arose as roles and responsibilities changed during and after the crisis and these are highlighted, as are the contrasting views of citizens and formal institutions on levels of tolerable risk.

It is important to point out that primary data was collected from the focus group discussions and interviews to supplement existing data and analysis of the Montserrat crisis and recovery processes. This explains the very limited number of interviews. While this has its limitations, the research team felt that governance during the crisis and its immediate aftermath had already been studied in depth, albeit from the perspective of science-policy interface (see, for example, Donovan *et al.* [2012]; Donovan and Oppenheimer [2013]). Further data was therefore collected to complement this and bring it up to date, situating the analysis of risk governance within broader decisions about development and the future of the island.

Disaster Risk Governance in Montserrat – an Unstable State

On the 18th July 1995, the Soufrière Hills Volcano became active after a long period of dormancy. Approximately 6,000 people were evacuated from the capital Plymouth and nearby towns to temporary shelters. They returned to their homes, were evacuated again, and on 3rd April 1996 Plymouth was evacuated for the last time. Approximately 1,300 people were housed in temporary public shelters, which suffered from overcrowding, lack of privacy, poor sanitation and lack of access to good nutrition. Many Montserratians left the island, supported by UK resettlement packages, family and friends. By 2001, the population of Montserrat had dropped by 60%, from 11,314 in 1991 to 4,491 in 2001 (CARICOM [2009]). For those that stayed, some were still in shelters three years after the eruption. Those that decided to stay and resettle in the north of the island, which is much drier and less fertile than the south and more exposed to hurricanes and flooding, faced severe challenges in re-establishing their livelihoods (Rozdilsky [2001]).

Re-settlement in the south meanwhile has been controlled and in some areas prohibited. Exclusion zones have been set up to control access to areas close to the volcano according to the level of volcanic activity (see Figure 2). These and other major risk management decisions are listed in Annex 1. The governance arrangements and relationships shaping these decisions and collective responses to volcanic risk are discussed below.

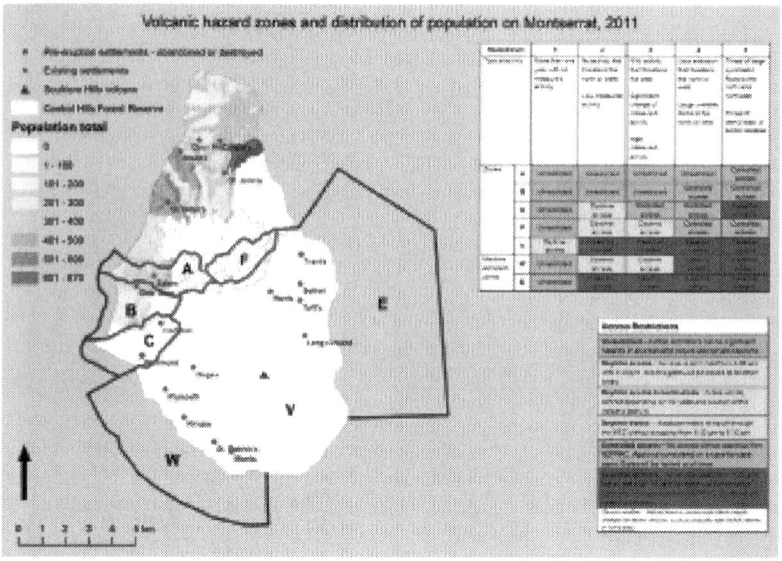

Figure 2: Map of exclusion zones, settlements in 2011 and pre-eruption settlements. Katy Mee, British Geological Survey.

Risk Governance before the Volcanic Crisis

Governance arrangements in UK overseas territories are unique because of their colonial history, although they have some similarities to structures found in decentralised systems of governance elsewhere. Local governments have autonomy over day-to-day decision-making and planning with regard to social and economic policy, receiving some budget support to do so, but defer to central government over decisions regarding internal security and defence. This includes emergency management functions, if the capacity of local government to respond is surpassed, but in pre- and post-disaster risk reduction decisions, local government is expected to play a dominant role.

From 1961 up until the volcanic crisis, the local government in Montserrat enjoyed very high levels of autonomy from the UK. The 1960s saw a period of decolonisation in the Caribbean and although Montserrat's leaders chose to remain part of Britain, the island became self-governing with the formation of a locally elected ministerial government. From then on, Montserrat, like the Turks and Caicos,

Cayman Islands and Anguilla, was treated as a quasi-independent state. A new constitution in 1989 set the parameters for these governance arrangements, giving the local government close to full autonomy over decision-making within the territory. The governor of Montserrat, a UK government representative and civil servant in the Foreign and Commonwealth Office (FCO), was responsible for defence, external affairs and internal security but performed mainly ceremonial roles. The local government meanwhile carried out most normal areas of government activity such as provision of health and education, policing and land-use planning with relatively little interference from the UK government, requiring minimal budget support and even developing some infrastructure projects independently (Clay *et al.* [1999]).

In terms of vertical coordination, a set of 'ad-hoc' and 'personalised' governance arrangements had evolved between the UK and its Caribbean Overseas Territories before the volcanic crisis. These reflected neither a sense of shared sovereignty (as in the French Caribbean) nor negotiated autonomy (as in the Dutch Caribbean), but rather an assumption by the UK government that these territories would become independent (Hintjens and Hodge [2012]: 202). Even the constitution created ambivalence, recognising Montserrat's separateness, but maintaining the UK's constitutional power to invoke emergency orders and intervene directly in domestic affairs.

In-line with this broad level of independence before the volcanic crisis, Montserrat was also free to design and implement its own policies in response to perceived disaster risks; however, limited local capacity to identify and analyse risk was only part of the problem. Concentration of political power within a few wealthy families, party politicking and personalised politics, common to other island states (Skinner [2002]) meant that policies were geared towards favouring interest groups not serving the needs of the most vulnerable.

Like many of its Caribbean neighbours, Montserrat is prone to a range of geological and hydro-meteorological hazards and yet risk management knowledge was not well developed and had not been incorporated into mainstream development (World Bank [2002]). Knowledge of volcanic risk was extremely low amongst local politicians and UK government representatives on island despite the publication of the Wadge and Isaacs report ([1986]), which had been commissioned by the Pan Caribbean Disaster Preparedness and

Prevention Project (CDPPP). The report warned of volcanic activity and the potential impact that an eruption would have to the island's capital, Plymouth. An early version of the report was discussed with the Permanent Secretary in the Chief Minister's office, yet there was no long-term planning for a volcanic eruption (Shepherd *et al.* [2002]). Many explanations have been offered for this omission, including a lack of previous experience with volcanic eruptions and the impenetrability of the scientific language, both of which meant that it was difficult for policy-makers to take the findings of the report seriously; as well as limited resources and the more immediate focus of dealing with hurricanes (interviews, local and UK government officials, Montserrat, 2–4 October 2012).

In 1989, Hurricane Hugo hit the island leaving 11 dead and over 3,000 homeless, as well as causing substantial damage to approximately 85 per cent of homes and to a number of the storm shelters (Berke and Wenger [1991]). A Hurricane Preparedness Scheme had been in place since 1980, but Hurricane Hugo revealed serious weaknesses in planning, including poor emergency shelter construction and lack of maintenance. Moreover, the risk control measures that were in place for this type of hazard, including local development regulations and inspection and enforcement procedures, had not been effectively implemented, and the housing stock was not designed using storm-resistant construction techniques. Unable to respond to the crisis with local resources, a state of emergency was declared and day-to-day control of the island passed away from the locally-elected Chief Minister to the FCO (Skinner [2006]: 57). The UK government took over emergency management efforts and the support was well received (£3 m in emergency aid and £16 m in long-term reconstruction) promoting a quick material recovery and allowing Montserrat to achieve a budgetary surplus by 1995 (Clay *et al.* [1999]).

Hurricane Hugo prompted a temporary alteration in the prevailing governance arrangements, with the local government losing decision-making autonomy and the UK becoming directly involved in local affairs. Montserrat is a contingent liability for the UK government, so when local capacity to respond was surpassed, the UK recognised its responsibility to intervene and assist the islanders (Hintjens and Hodge [2012]). Lack of planning and heavy dependency of foreign assistance led to a 'loss of control on the part of Montserrat authorities' (Berke and Wenger [1991]: 77), but this was not permanent and six months after the hurricane, Montserratian authorities were exerting substantial control over the recovery process and development plans.

Abrupt social events allow hitherto marginalised issues to get on the agenda, by opening up 'policy windows' and creating spaces for policy reform (Kingdon [1995]). In the same way, major disasters can act as 'focusing events' by bringing the failures of existing disaster policies to the attention of the public and policy makers, opening up policy windows for DRM reform (Birkland [2006]). Hurricane Hugo made it clear to local authorities that a more coordinated effort was needed to prepare for and respond to extreme events and in 1994 a National Disaster Action Plan was drawn up and an Emergency Operations Centre (EOC) established in 1995. However, for the reasons described above a volcanic eruption was not on the political radar either for inclusion in the plan or reconstruction efforts after Hurricane Hugo. Indeed, the £16 million investment in reconstructing Plymouth, building a new hospital and housing, would have acted as a major disincentive to investing elsewhere even if volcanic risk had been taken seriously. As such, reducing risk to hurricanes in the post-disaster reconstruction efforts locked Montserrat in to high exposure to volcanic risk and a development trajectory that would prove difficult to alter in the face of an abrupt change in the volcanic hazard.

Prior to the volcanic crisis Monserrat was poorly integrated in regional and international risk governance systems. There was no formal mechanism through which Montserratian authorities could access resources or advice on disaster scenarios, potential impact and risk reduction options, although in fairness the international community as a whole understood little about the social or political sources of disaster risk in 1995. International and regional organisations at that time were promoting scientific, engineering and bureaucratic (or 'technocratic') solutions to disaster problems (Hewitt [1995]; Cannon [1994]). The Caribbean Disaster Emergency Response Agency (CDERA), set up in under the Caribbean Community and Common Market (CARICOM) in 1991, was, as its name suggests, a response-focussed agency with objectives of coordinating relief efforts, channelling aid from NGOs and other governments, mitigating the immediate consequences of disaster and improving disaster response capacity amongst participating states. As such, it provided little guidance as to how to assess and manage risk. Montserrat could not expect much in the way of technical support or guidance from the UK government either, as it did not have a DRM plan of its own at that time - the Civil Contingencies Act was not brought into force until November 2005. Overall, the lack of coherence across knowledge systems, resulted in a limited consideration of any hazards in development policies and plans. In particular, it made Montserrat highly susceptible to the unknown risks associated with the Soufrière Hills volcano.

During the Crisis

Emergency management during the crisis has been characterised as unplanned, reactive and short term (Clay *et al.* [1999]). Lack of preparedness meant that 'actions taken by the UK Government and the Government of Montserrat were driven stepwise by events in the volcanic escalation' (Kokelaar[2002]: 1). Unlike Hurricane Hugo, where Montserrat's independence from the UK remained largely unaltered despite huge investments in reconstruction, the volcanic crisis brought about an abrupt realignment of vertical governance arrangements, with the UK government's position towards this overseas territory shifting radically towards greater intervention as the crisis unfolded. Even as Montserrat moves beyond recovery into processes of longer-term development, central-local relations have not returned to their pre-eruption state.

The EOC was the key local government entity managing the response to the volcanic eruption (Clay*et al.* [1999]). Though a nominally 'local' institution led by the chief minister's office, the EOC is activated by the governor who on 3 April 1996 declared a state of emergency, thus rendering the EOC subservient to the governor's office and ultimately the FCO. At the beginning of the crisis the EOC made some decisions about planning and coordination of evacuations, supplies and shelters; but once the state of emergency was declared it no longer made any substantive decisions without the governor's consent. In small face-to-face societies 'people take on a number of roles and might interact with each other in different capacities at different times of the day [and] [t]his can make communication very difficult' (Skinner [2002]: 307). During the crisis these norms of communication were suddenly altered by changes in the already complex functions of different actors, often creating tension – for example between the chief minister and the governor.

In addition to this shift in decision-making authority, the capacity of the EOC to make decisions regarding emergency response was tested and found wanting, as decisions during the crisis became more complex. In shelter management, for example, the EOC had no special expertise or sensitivity to the importance of engaging people in decisions (Clay *et al.* [1999]: 70). Moreover, as people (and particularly the middle class) began to leave the island as the crisis intensified, local management capacity was further eroded.

The volcanic crisis was marked by a lack of contingency planning or strategy for how the FCO and the then the Overseas Development Agency (ODA) would manage a complex and long duration emergency in an overseas territory: 'Ad-hoc arrangements had to be put in place and this was done reactively as the eruption progressed' (Clay *et al.*[1999]). The strategy adopted was to react to changing hazard levels as they were identified but this lack of planning, coupled with low levels of communication and community consultation, meant that UK and 'local' ideas about how to manage emergency response often diverged.

Weaknesses in planning were also partly due to poor horizontal coordination between the FCO, which delegated advice on external affairs, civil order and financial matters to the Dependent Territories Regional Secretariat (DTRS) in Barbados, set up in 1993, and ODA. Each had responsibilities and roles to play in an emergency situation but there were some unclear areas of responsibility within this complex set of horizontal institutional arrangements, resulting in a fragmentation in authority (Clay *et al.* [1999]: 16). Prior to the crisis, Montserratian authorities had become accustomed to dealing only with the DTRS but as the crisis evolved, other departments and individuals would become more directly involved in emergency aid, splitting the decision-making responsibility and resources across branches of government. This 'bizarre situation', as it was referred to by journalist Polly Pattullo ([2000]: 137), was compounded by insufficient mechanisms for inter-departmental coordination of responsibilities in London (Clay *et al.* [1999]: 16). In addition, aid coordination was complicated by donations coming in from a range of sources including bilateral aid from CARICOM countries, regional/multilateral aid from the Caribbean Development Bank (CDB), the European Commission Humanitarian Office (ECHO), and from NGOs. Montserrat was not short of emergency relief, according to local residents, but there were not enough trained people to handle it and this, along with delays at customs because packages were not properly labelled, slowed down the process (focus group discussions, 27 September 2012).

Trust between the state and society can be caused and aggravated by low levels of formal public consultation on – as well as public willingness to participate in - decisions regarding emergency management (Wilkinson [2012]). The emergency aid programme in Montserrat was implemented with little local consultation creating tensions between UK and local authorities, a deepening sense of insecurity amongst residents and growing mistrust between local stakeholders and UK government. As tents, cots and army rations were distributed, the inappropriateness of many of the supplies became apparent (interviews and focus groups, various, 28 September - 3 October 2013). Examples included bringing in pit latrines, which had never before been used on the island, and tents to be used as shelters, which would not withstand tropical storms and were inappropriate for the heat; all of which could have been avoided by consulting local authorities. Conversely, although citizens were able to express their views on both local and UK government handling of affairs through radio programmes, they were reluctant to go to town meetings. Hence, formal channels of social participation in decisions-making were very limited (interviews, local government officials and residents, 1 and 3 October 2012).

Lack of coherence between local and UK authorities over policy direction also contributed to growing mistrust. The local government preferred a 'wait and see' approach during the early phases of the emergency, assuming less serious impacts from the eruption, which resulted in deferral of UK-funded public housing construction in the north. The UK government, on the other hand, preferred to plan for the worst case, because of its ultimate responsibility for Montserrat (Clay et al., [1999]: 54). This included drawing up a plan for the complete evacuation of the island, known as Operation Exodus. Operation Exodus had existed since the early days, but did not become public knowledge until May 1998, which generated rumours of 'relocation schemes' and plans by the UK government to 'de-populate the island' (interviews, local residents, 3 October 2012). The UK government was unlikely to have had any genuine desire to empty the island but the lack of a public communications strategy on shelters, evacuations and recovery plans had negative repercussions with Montserratians commonly expressing the view that 'the UK Government wanted us off the island' (interview, local resident, 3 October 2012).

Coherence in emergency management was complicated by the various vertical lines of communication that existed between different UK departments and local authorities and between the scientists on and off the island and UK and local authorities. In particular, the volcanic crisis highlighted the lack of local capacity to translate and communicate scientific information and this had repercussions for awareness of risk amongst local government officials and the public:

There was not a systematic analysis of scientific advice and policy-makers did not know what questions to ask... The Wadge Report was a perfect example of that: no one took any notice because it was not translated into practical advice (interview, UK government official, 2 October 2012).

From a local government perspective, clearer messages were needed and expected to help interpret volcanic hazard information, as one local government official explained (interview, local government official, 4 October 2012):

We had little experience with scientists. With hurricanes they are more hands off; they can show you on a computer and it is easier to understand. With a volcano it is difficult to see anything on which to base a decision, plus the scientists kept saying 'this is not an exact science'. In an effort to be cautious they actually reduced their own credibility and the public started to doubt.

The failure to articulate and coordinate policy direction also delayed reconstruction efforts and crucially, the decision to invest in the north and hence drastically reduce levels of volcanic risk on the island. The UK had put money into rebuilding Plymouth and continued to see it as the island's capital, and for this reason the Department for International Development (DFID) was reluctant to start buying up land in the north and building houses there (Clay *et al.*[1999]). More broadly, the UK government was waiting for the volcano to stabilise before re-investing in the island's infrastructure, and at the same time, Montserratian authorities wanted to avoid sending out the wrong signals and were keen to maintain a 'business as usual' atmosphere to keep people on the island and keep the economy going (Skinner [2002]). This may explain why it did not put more pressure on the UK government or ask for money to start building in the north; but the result was that two years after destruction of Plymouth, over 300 people were still living in temporary shelters (Haynes *et al.*[2008]; Skelton [2003]).

Despite complex organisational structures and unclear mandates, coordination of emergency management did improve as the crisis progressed. The administration of shelters improved for example when the UK government responded to complaints about aid in 1996 by introducing a food voucher scheme. In 1997, the vouchers were replaced by cheques as a pragmatic response to pressure for more flexibility, so people could use the income to pay other expenses such as rent (focus groups, various, 28 September - 3 October 2013). This also reduced the heavy administrative burden of the voucher scheme (Clay *et al.*[1999]).

After the Volcanic Crisis

It is difficult to identify the exact point at which emergency management ended and longer-term recovery planning began, as recovery has not been a geographically evenly distributed phenomenon, with 'different areas of the island... in different stages of the recovery process' (Rozdilsky [1999]: 6). Similarly, it is hard to identify the time when the people of Montserrat accepted and began to plan for a new future in the north of the island. Certainly, the 19 deaths on 25 June 1997 were 'a game changer' (interview, UK government official, 2 October 2012).

A major turning point relates to the type of support Montserrat received from the UK. From late 1997 onwards, emergency aid was increasingly outweighed by budget support and substantial capital investments to re-establish basic services, develop infrastructure and provide incentives and an enabling environment for private investment and longer-term development. From 1997 to 2012, DFID spent £325 million on technical assistance, budgetary support and capital investments, representing 50 per cent of the total spent on Overseas Territories during that period. Six capital investment projects alone (an airport, roads, water, power and education) involved an investment of over £34 million (ICAI [2013]). However, the scale of this investment came at the price of heavy reliance on the UK, and although the local government is keen to avoid long-term dependency and achieve self-sufficiency but there is no realistic plan for doing so. The 2011 Strategic Growth Plan, for example, creates 'no overall picture of self-sufficiency for the island' (ICAI [2013]: 8). A reliance on the UK for capital is compounded by the fact that Montserrat cannot access development finance from other sources. It is not eligible for loans from the World Bank or International Monetary Fund, although it receives some funds from the EU and CDB - £4.8 million from 2012 to 2015, but this merely supplements the £24 million committed by DFID for the same period (ICAI [2013]).

A change of government in the UK in May 1997 had far-reaching consequences for risk management in Montserrat, with the UK government at senior level taking more of an interest. The Montserrat Action Group was formed and the then Secretary of State for International Development Claire Short established a joint DFID-FCO review of off- and on-island options, and £6.5 m was allocated by the UK government for development in the north. Coordination of recovery efforts improved thanks to a clarification of mandates in London within one department - the Conflict and Humanitarian Affairs Department of DFID – which was made responsible for coordination of all financial aid and equipment to Montserrat. However, this had the impact of separating UK development and foreign policy, with Montserrat's governor reporting to the FCO and the Aid Office reporting to DFID, essentially separating safety from funding.

Despite its financial dependency on the UK, improvements in DRM policy and organisational structures owe more to Montserrat's insertion in the regional disaster risk governance system. The new unit in government in 1997, now called the Disaster Risk Management Coordination Agency (DMCA), set up to coordinate DRM activities, and the Disaster Preparedness and Response Act of 1999, were based more on examples from around the Caribbean than the UK disaster management system. CDERA (which later became the Caribbean Disaster Emergency Management Agency (CDEMA) in 2009) adopted a comprehensive disaster management approach and national emergency management offices across the Caribbean have followed suit. These strategies also reflect the language and priority areas of the Hyogo Framework for Action 2005–2015; demonstrating the influence that international policy has had on regional risk management.

Integration in the regional risk governance system deepened in 1999, when the Montserrat Volcano Observatory Act was passed, bringing it under local legislation and encouraging 'collaborative links with regional and extra-regional centres of expertise in scientific disciplines relevant to monitoring volcanic activity' (1999, Art.8). It was now seen as a locally owned institution (interview, UK government official, 2 October 2012). Montserrat also began to receive advice and support from CDEMA, as one of 18 participating states and was included in the Action Plan 2011–2012 for the Caribbean, promoted by the disaster preparedness programme of the European Commission›s Humanitarian Aid and Civil Protection Directorate General (DIPECHO). Montserrat's Sustainable Development Plan 2008–2020 now reflects a comprehensive disaster management mentality, with a Strategic Goal on environmental management and disaster mitigation that emphasises governance structures, training and education on DRM and building response capacity at all levels.

Even more encouragingly DRM is now seen to be an integral part of the development process, at least on paper. Local government authorities recognise that disasters can lead to major disruptions to the island's developmental agenda (Government of Montserrat [2005]). The Montserrat Corporate Plan 2003–2006, health, water and education sector plans all included DRM elements, although these mainly focussed on streamlining disaster preparedness and response. In 2003, for example, the Ash Clearing Assistance Project concentrated on reducing air pollution and health hazards in the environment after the volcano dome collapse. Local ownership over emergency response was clearly demonstrated at this time: the Montserrat government declared the disaster and activated the emergency operations centre, which then acted as the coordination body for response and relief efforts.

Notwithstanding these improvements however, decision-making authority on island and the coordination of DRM activities is still limited by the absence of an inclusive DRM plan. As of October 2012, the Disaster Management Plan was still not finalised and had only been updated in an ad-hoc manner by the DMCA director. Hence the content of the plan and allocation of responsibilities remains unclear to other government officials (interviews, local government officials, 3–4 October 2012). The DMCA is an operational not regulatory agency with a mandate to prepare for emergencies, not reduce levels of risk in society and so can only play a limited role in strengthening local DRM capacity on-island. As one local government official commented:

Institutions are stronger, but high staff turnover and lack of technical experience mean that an effective disaster response in the future will require quick funding and external support. The Government of Montserrat will be able to respond in a limited way for a week or two but will need financial support and technical assistance (interview, local government official, 3 October 2012).

By the end of 1997 the north was deemed safe for occupation but people were still living in shelters. A Sustainable Development Plan was produced identifying health, education and housing investments needed for economic and social recovery, but many of these investments were undertaken with only a short-term focus: the hospital was upgraded at the St John's site, not rebuilt; an emergency jetty was built at Little Bay instead of a harbour; and only a temporary government headquarters were set up in Brades (Sword-Daniels *et al.*[2013]). For many, it was not until 1999 that the emergency phase really ended. Eruptive activity continued, but a new governance regime was beginning to emerge with a vision of the island's future development. This 'co-governance' regime would continue to dominate central-local relations in Montserrat to the present day. The local government began to take the lead on day-to-day management functions, such as the procurement and management of development projects and some control over spending decisions, but with strong oversight and financial control from the UK. Montserrat regained some autonomy with respect to the crisis period, but compared with the pre-crisis situation, economic dependency remained high:

DFID keep changing the rules of the game, including greater scrutiny of expenditures, increasing limits to what officers can approve now (compared to 1995). All this affects our ability to respond quickly to people's needs (interview, UK government official, 3 October 2012).

Greater coherence between UK and local risk perceptions and DRM activities can be observed from 2001 onwards with the development of a strategy to sustain the on-island community and promote long-term investment in the north of the island (Clay *et al.*[1999]: 13). By restricting access to proximal areas (the boundaries of which have changed over time (Aspinall *et al.*[2002])) and investing in basic and road infrastructure, housing and services in the north, levels of exposure to pyroclastic flows and lahars have all been dramatically reduced (Sword-Daniels *et al.*[2013]). For the Montserrat Government these decisions marked an important turning point in the recovery process:

In 2001 the economy began to recover and economic plans were made, based on scientific advice. The Scientists said that the far north was of low negligible risk. Once that was said they set the foundations for serious thinking about investment for those who stayed. They realised it would have to be in the north (interview, local government official, 3 October 2012).

There was no formal public consultation process to establish how different actors viewed volcanic risk on the island (Haynes *et al.* [2008]), however, perceptions of risk appear to have been broadly aligned at this point with residents beginning to consider the north their permanent home (interviews, local residents, 1–3 October 2012). Many had already left the island after by the Boxing Day collapse in 1997 and facilitated by relocation packages offered in 1998, but even for those that stayed and had lost their houses, land and jobs, the north did not represent an 'acceptable option' in terms of levels of risk and livelihood options until housing reconstruction began (interview, UK government official, 2 October 2012). This perception of the south being dangerous (approximately 60 percent of the island) and the north being safe for habitation was broadly in line with the scientific assessments, through which areas were established as exclusion zones – some permanently, and others in accordance with the level of volcanic activity. Despite informal reports of people entering the permanent exclusion area (Zone V) without permission, the current general perception amongst islanders is that this area will continue to be highly exposed to volcanic hazards and they will never be able to return (interviews and focus groups, various, 28 September - 3 October 2013).

In the transition from recovery to longer-term development, greater vertical coherence in development planning has emerged. This owes much to harmonisation across departments in the UK, with ODA being upgraded to ministry status and re-named DFID, with overall responsibility for the aid budget. A team of programme officers for Montserrat was created within DFID and on island (a resident lead, an infrastructure adviser and two programme officers) (ICAI [2013]). Forced on to the political agenda in the UK by the volcanic crisis, these changes –outlined in the 1999 White Paper*Partnership for Progress and Prosperity* and 2002 British Overseas Territories Act– have had important implications for inter-governmental mandates: any laws adopted by the UK or through the European Union, are now applicable to Montserrat. This includes more stringent EU environmental laws. The Act has brought about increased consultation between political cadres of territories and the UK government, and a more proactive dialogue has opened up (interviews, UK and local government officials, 1–2 October 2012).

The volcanic crisis had uncovered some of the inherent contradictions in the autonomous system of governance in Montserrat, but it also prompted UK and local authorities to consider their own levels of risk tolerance and responsibilities for reducing exposure. A tacit alignment of UK and local government risk perceptions was established in 1999 and endorsed through the definition of an exclusion zone and subsequent investment in the north. Yet more recent studies suggest that these formal notions of tolerable risk may not be shared by the local population or even the local government (Haynes *et al.* [2008]). Exposure to high-impact events such as pyroclastic flows has been dramatically reduced through officially prescribed norms intended to reduce risk (relocation and establishing exclusion zones), but the north of the island had been affected by ash fall and acid rain until recenly, representing a low-level, intermittent but widespread risk that is given low consideration in development planning. Ash fall presents health risks and asthma sufferers in particular have reported suffering respiratory problems from heavy ash fall (interviews, local residents, 1 to 3 October 2012). Infrastructure built during the recovery period has also been affected and needs constant cleaning, replacement and repair. Most buildings (and homes) have tropical slatted windows, which allow ash to enter buildings because they cannot be properly sealed (Sword-Daniels *et al.*, [2013]).

The negative impacts of relocating people in the north of the island and of the social upheaval of Montserratians moving to the UK should not be overlooked. Most Montserratians on the island today, are worse off economically that before the eruption. Farming activities are less lucrative and farmers are reluctant to invest as they do not have security of tenure and are aware of the threat of future ash fall and acid rain (Halcrow Group and the Montserrat National Assessment Team [2012]). Land shortages in the north have meant that new houses have been erected in unsafe and unsuitable locations such as ravines (Hicks and Few, [2014]). Although resettlement has reduced exposure to volcanic hazards, these policies have created new vulnerabilities for the island population that may be more tolerable than volcanic hazard exposure for now, but this may not always be the case.

There are instances of individuals not subscribing to official rules, which suggests that levels of risk tolerance vary and are not static. People entering the exclusion zone for livelihood reasons, such as tending to crops and illegal scrap metal collecting, as well as those building too close to the exclusion zone, are examples of this. Expatriate residents continue to live in Old Towne, which can become part of the exclusion zone with heightened volcanic activity, and have expressed their reluctance to evacuate and lack of confidence in the alert levels issued by the MVO and temporary evacuations decisions (interviews, local residents, 3 October 2012). Nearby Salem has a secondary school and a primary healthcare clinic and is home to a growing immigrant population and an ad-hoc business district (Sword-Daniels *et al.* [2013]). Rental housing is cheaper in this area and new arrivals appear to be less aware of the risks associated with volcanic activity than Montserratians (interviews, local residents and local government, 1 to 3 October 2012). Overall, the view that the future of the island is in the north appears not be as unanimous as official views and recent patterns of infrastructure investments suggest.

Another factor suggesting that local authorities may not entirely endorse the idea of development in the north is the temporary nature of much of the island's vital infrastructure. Sword-Daniels *et al.* ([2013]) note that many of the buildings and essential services that were put up during the recovery period were not permanent structures. These facilities have been upgraded incrementally over time but the perception of sites as temporary has in some cases obstructed funding leaving some buildings in an inadequate state. These 'quick fixes' need to be re-addressed to enable further progress towards development goals.

DISCUSSION

The disaster risk governance regime in Montserrat has undergone a radical shift as a result of the volcanic crisis of 1995–1997 and alterations in central and local perceptions of volcanic risk. Essentially, a longer-term view of risk has been adopted by UK and local authorities, scientists and local communities, and this has brought with it substantial investments in safer locations further north and a belief that the future of the island is in the north. DRM in Montserrat is no longer concerned with the circumstances under which a return to south will be possible or how to make lives and livelihoods safer in former settlements. The longer term view of risk management being taken and new investments being made in safer locations further from the volcano, represents an important shift in the risk governance system (see Table 2).

Table 2: Nature and processes of disaster risk governance in Montserrat

Change in disaster risk governance system		Phases of volcanic risk		
		Pre-crisis	During crisis	Post-crisis
Nature of change	(a) Dispersion of decision-making	(a) Little UK interference in development decisions.	(b) Scaling-up of UK intervention in response to weak local capacity.	(a) New co-governance regime with low levels of local autonomy.
	(b) Vertical coherence	(b) Poorly articulated DRM policies and little coherence.	(b) Problems with policy coherence creating confusion over policy aims, lack of confidence in government and mistrust.	(b) Greater vertical coherence on DRM – e.g. linking up with other scales (regional and EU).
Change process	(a) Stability	(a) Temporary shift to high levels of external control (after Hurricane Hugo).	(a) Temporary shift to high levels of external control.	(a) Sustained shift to high levels of external control. Irreversible shift to greater vertical coherence.

	(b) Extent	(b) Little change.	(b) Incremental change to vertical relations during crisis.	(b) Transformation in decision-making control. Incomplete transformation in vertical coherence.

Wilkinson

Wilkinson *Journal of Applied Volcanology* 2015 4:3, doi:10.1186/s13617-014-0021-7

A transformation towards greater vertical coherence has also taken place but is not complete, and there are signs that local and external-scientific assessments of volcanic risk in Montserrat are diverging. In particular, scientists and UK government officials have raised concerns about increasing settlement in areas close to the exclusion zone plus the low consideration given to ash fall in development planning (interviews, UK government officials and scientists, Montserrat, 2–4 October 2012). These comments and trends collectively suggest that the tolerable level of risk for local residents is higher in some cases than that established by UK and Montserratian authorities. Similarly, international development agencies have expressed concern that public awareness of hazards other than volcanoes needs to be improved. According to a review of disaster risk management capacity in Montserrat carried out by UNDP ([2010]), the focus of DRM activities is too often related to the Soufrière Hills volcano, with insufficient emphasis on a multi-hazard approach.

These local perceptions of risk and the cognitive processes through which risks are deemed insignificant or adequately controlled by individuals and groups need to be explored further and contrasted with external and scientific judgements. Calculations of tolerable risk are not static and the analysis presented above demonstrates how both new people coming into a volcanic area and the passing of time may change 'local' perceptions of risk. The Montserrat case does however suggest that transformational shifts in disaster risk governance can only occur when tolerable levels of risk are agreed on by stakeholders and this will require high levels of horizontal as well as vertical and coherence.

In analysing continuities and discontinuities in Montserrat's disaster risk governance system from the late 1980s to the present day, alterations in the governance system can be observed on two occasions: in the aftermath of Hurricane Hugo and during the volcanic crisis period. For both events, abrupt changes in levels of disaster risk and limited local capacity to respond led to greater external interference in local DRM decisions. Although Hurricane Hugo was a high impact event, the hazard subsided quickly and these alterations were temporary. The volcanic eruption, on the other hand, occurred over a long period of time and produced more permanent changes in the disaster risk governance regime and in the island's governance system more broadly. The sharp and sustained rise in the level of volcanic risk combined with a weak response from local and UK authorities led to a sustained reduction in local autonomy but also an increase in vertical coherence and when levels of risk declined and post-disaster recovery ended these new configurations did not return to their pre-crisis state.

This transformation may not prove to be irreversible, although there could be a latent 'tendency towards dependency' in Montserrat common to all UK overseas territories (Pattullo [2000]; Skinner [2002]). For critics of UK colonialism these territories 'will always struggle to develop and will always be dependent upon other places and people' (Skinner [2002]: 316). One aspect of the shift in risk governance in particular that may be permanent is that of increased vertical coherence. Although local capacity to assess risk and implement risk reduction measures is still limited by lack of human and technical resources, Montserrat is now better integrated into a regional disaster risk governance system that can offer this support and advice. This is unlikely to change.

CONCLUSIONS

This research draws a number of conclusions about volcanic crises and regime change in Montserrat of relevance to multi-tiered governance regimes elsewhere and to different hazardous contexts. The examples of Hurricane Hugo and Soufrière Hills both suggest that crises brought about by sharp increases in the level of risk are likely to provoke temporary alterations in central-local relations, and in particular a sharp decline in local autonomy over DRM decisions. This intervention by external actors can have both negative and positive consequences for disaster risk management, creating dependency but also enhancing vertical coherence, offering opportunities for learning and capacity building.

The Montserrat experience is atypical however and caution should be exercised in drawing lessons for other contexts. In particular, the relationship between the UK and its overseas territories is unique and different even from French and Dutch overseas territories in the Caribbean. Central governments elsewhere may not be so inclined to provide ongoing financial support to local governments after the recovery process is considered to have ended. Similarly, local governments with significant levels of autonomy in decentralised and particularly federal systems of governance elsewhere are likely to reject sustained central government interference in local affairs following a protracted crisis. Governance reform in Montserrat was the product of conflict, but ultimately compromise, and in other contexts consensus between central and local authorities on tolerable levels of risk may be harder to achieve. Notwithstanding these caveats, however, the transition to co-governance and the re-framing of disaster risk that have taken place in Montserrat provide useful examples of how transformations can occur in disaster risk governance systems following high-intensity, long-duration volcanic events.

The experience of Montserrat also provides useful insights for volcanic islands elsewhere and small island states with disaster risks more generally. Small islands have few options for resettlement when significant parts of the territory are destroyed by a disaster, or when the decision is taken to move populations before a disaster to prevent loss of life. The benefits in terms of reducing disaster risk have to be weighed against loss of livelihoods for a significant proportion of the population, considerable social upheaval and often economic decline. Critical to the success and sustainability of these risk management decisions is the need for vertical coherence and dialogue between different scales of governance. In Montserrat this has been partly achieved through greater integration into the regional risk governance system and via the establishment of an economically dependent but politically autonomous system of co-governance with the UK. But unless communities are also engaged in risk governance decisions and consensus is built, this tacit agreement to pursue a low-volcanic-risk development model could come unstuck. Small islands with large risks can learn from the Montserrat experience. They can anticipate and plan for how these dialogues might take place in the event of a major disaster.

ACKNOWLEDGEMENTS

The author would like to thank Edward Clay for early discussions on UK-Montserrat relations, Richie Robertson, Barbara Carby and Peter Simmons for providing comments and suggestions on how to improve this paper and to three anonymous reviews who provided useful comments. Research was conducted under the STREVA project, funded by the NERC/ESRC *Increasing Resilience to Natural Hazards in Earthquake-prone & Volcanic Regions* programme.

REFERENCES

1. Alexander D (2002) Principles of Emergency Planning and Management. OUP, New York.
2. Argyris C (1999) On Organizational Learning. Blackwell, Oxford.
3. Aspinall WP, Loughlin SC, Michael FV, Miller AD, Norton GE, Rowley KC, Sparks RSJ, Young SR (2002) The Montserrat Volcano Observatory: Its Evolution, Organization, Role And Activities. In: Druitt TH, Kokelaar BP (eds) The eruption of Soufrière Hills Volcano, Montserrat, from 1995 to 1999, Geological Society, London, Memoirs.. pp 71-91
4. Barnes P (2002) Approaches to community safety: risk perception and social meaning. Austr J Emerg Manage Autumn 2002:15-23
5. Bené C, Godfrey Wood R, Newsham A, Davies M (2012) Resilience: new utopia or new tyranny? Reflection about the Potentials and Limits of the Concept of Resilience in Relation to Vulnerability Reduction Programmes. IDS Working Paper 2012 (405). Institute of Development Studies, Brighton.
6. Berke P, Wenger D (1991) Montserrat: Emergency Planning, Response and Recovery Related to Hurricane Hugo. Texas A&M University, Hazard Reduction and Recovery Center, College Station, Texas.

7. Birkland TA (2006) Lessons of Disaster: Policy Change after Catastrophic Events, American Governance and Public Policy Series. Georgetown University Press, Washington DC.
8. Cannon T (1994) Vulnerability Analysis and The Explanation Of 'Natural' Disasters. In: Varley A (ed) Disasters, Development And Environment. John Wiley, Chichester. pp 13-30
9. (2009) National Census Report. Montserrat, CARICOM Secretariat, Georgetown, Guyana.
10. Christoplos I, Mitchell J, Liljelund A (2001) Reframing risk: the changing context of disaster mitigation and preparedness. Disasters 25(3):185-198
11. Clay EB, Barrow C, Benson C, Dempster J, Kokelaar BP, Pillai N, Seaman J (1999) An Evaluation of HMG's Response to the Montserrat Volcanic Emergency. Department for International Development, London.
12. Cutter S, Barnes L, Berry M, Burton C, Evans E, Tate E, Webb J (2008) A place-based model for understanding community resilience to natural disasters. Glob Environ Chang 18(4):598-606
13. Donovan A, Oppenheimer C (2013) Science, policy and place in volcanic disasters: insights from Montserrat. Environ Sci Policy 39:150-161
14. Donovan AR, Bravo M, Oppenheimer C (2012) Co-production of an institution: Montserrat Volcano Observatory and social dependence on science. Sci Public Policy 40(2):171-186
15. Gaillard JC (2008) Alternative paradigms of volcanic risk perception: the case of Mt. Pinatubo in the Philippines. J Volcanol Geotherm Res 172(3–4):315-328
16. Gibson C, Ostrom E, Ahn TK (2000) The concept of scale and the human dimensions of global change: a survey. Ecol Econ 32:217-239
17. (2005) Montserrat National Report And Information On Disaster Reduction. Paper presented at the World Conference on Disaster Reduction. Kobe, Japan.
18. Gregg CE, Houghton BF, Johnston DM, Paton D, Swanson DA (2004) The perception of volcanic risk in Kona communities from Mauna Loa and Hualalai volcanoes, Hawaï. J Volcanol Geotherm Res 130(3–4):179-196

19. Haynes K, Barclay J, Pidgeon N (2008) Whose reality counts? Factors affecting the perception of volcanic risk. J Volcanol Geotherm Res 172:259-272
20. Heijmans A (2001) Vulnerability: A Matter of Perception. Disaster Management Working Paper 4/2001. Benfield Greig Hazard Research Centre, UCL, London.
21. Heinelt H (2002) Achieving Sustainable And Innovative Policies Through Participatory Governance In A Multi-Level Context: Theoretical Issues. In: Heinelt H, Getimis P, Kafkalas G, Smith R, Swyngedouw E (eds) Participatory Governance In Multi-Level Context: Concepts And Experience, Leske + Budrich, Opladen. pp 17-32
22. (2012) Participatory Poverty Assessments. Halcrow Group Limited, London.
23. Hewitt K (1983) The Idea Of Calamity In A Technocratic Age. In: Hewitt K (ed) Interpretations Of Calamity: From The Viewpoint Of Human Ecology, Allen and Unwin, Boston MA. pp 3-32
24. Hewitt K (1995) Regions of Risk: A Geographical Introduction to Disasters. Longman, Harrow.
25. Hicks A and Few R (2014) Trajectories of social vulnerability during the Soufrière Hills volcanic crisis. Journal of Applied Volcanology (in press)
26. Hintjens H, Hodge D (2012) The UK Caribbean Overseas Territories: governing unruliness amidst the extra-territorial EU. Commonwealth Comp Polit 50(2):190-225
27. (2013) DFID's Support To Capital Projects in Montserrat. Independent Commission for Aid Impact, London.
28. Kingdon J (1995) Agendas, Alternatives and Public Policies. Harper Collins, New York.
29. Kokelaar BP (2002) Setting, chronology and consequences of the eruption of Soufrière Hills Volcano, Montserrat (1995–1999). Geol Soc London Memoirs 21:1-43
30. McGinnis M (2000) Polycentric games and institutions: readings from the Workshop in Political Theory and Policy Analysis. University of Michigan Press, Ann Arbor MI.
31. Ostrom V (1997) The meaning of democracy and the vulnerability of democracies. University of Michigan Press, Ann Arbor MI.

32. Ostrom E, Burger J, Field CB, Norgaard RB, Policansky D (1999) Revisiting the commons: local lessons, global challenges. Science 284:278-282
33. Pahl-Wostl C (2009) A conceptual framework for analysing adaptive capacity and multi-level learning processes in resource governance regimes. Glob Environ Chang 19:354-365
34. Paton D, Millar M, Johnston D (2001) Community resilience to volcanic hazard consequences. Nat Hazards 24(2):157-169
35. Paton D, Sagala S, Okada N, Jang LJ, Burgelt PT, Gregg CE (2010) Making sense of natural hazard mitigation: personal, social and cultural influences. Environ Hazards 9(2):183-196
36. Pattullo P (2000) Fire From The Mountain: The Tragedy Of Montserrat And The Betrayal Of Its People. Constable, London.
37. Pelling M (2011) Adaptation To Climate Change: From Resilience To Transformation. Routledge, London.
38. Pidgeon N, Hood C, Jones D, Turner B, Gibson R (1992) Risk Perception. In: Risk Analysis, Perception And Management: Report Of A Royal Society Study Group. The Royal Society, London. pp 89-134
39. Posner PL (1998) The Politics of Unfunded Mandates: Whither Federalism?. Georgetown University Press, Washington DC.
40. Ramalingam B, Jones H, Reba T, Young J (2008) Exploring The Science Of Complexity: Ideas And Implications For Development And Humanitarian Efforts, ODI Working Paper 285. Overseas Development Institute, London.
41. Renn O (2008) Risk Governance: Coping With Uncertainty In A Complex World. Earthscan, London.
42. Rozdilsky JL (1999) Disaster Recovery In An On-Going Hazard Situation On Montserrat: the July 20, 1999, Volcanic Dome Collapse. Department of Resource Development Urban Studies Michigan State University, Michigan.
43. Rozdilsky JL (2001) Second hazards assessment and sustainable hazards mitigation: disaster recovery on Montserrat. Nat Hazards Rev 2001:64-71
44. (2010) The Hidden Dynamics Of Path Dependence: Institutions And Organizations. Palgrave Macmillan, Houndmills.

45. Shepherd JB, Robertson R, Latchman J, Lynch L (2002) Precursory Activity to the 1995 eruption of the Soufrière Hills Volcano, Montserrat. Seismic Research Unit, The University of the West Indies, St. Augustine, Trinidad.
46. Simmons P, Walker G (1999) Tolerating risk: policy principles and public perceptions. Risk Decis Pol 4(3):179-190
47. Skelton T (2003) Globalizing Forces And Natural Disaster: What Can Be The Future For The Small Caribbean Island Of Montserrat? In: Kofman E, Young G (eds) Globalization: Theory And Practice, 3rd edn. Continuum, London.
48. Skinner J (2002) British constructions with constitutions: the formal and informal nature of 'Island' relations on Montserrat and Gibraltar. Soc Identities J Stud Race Nation Cult 8(2):301-320
49. Skinner J (2006) Disaster Creation In The Caribbean And Planning, Policy And Participation Reconsidered. In: Pugh J, Henshall Momsen J (eds) Environmental Planning In The Caribbean, Ashgate Publishing Limited, Aldershot. pp 53-72
50. Slovic P (1987) Perception of Risk. Science 236:280-285
51. Sword-Daniels V, Wilson TM, Sargeant S, Rossetto T, Twigg J, Johnston DM, Loughlin SC, Cole PD (2013) Consequences Of Long-Term Volcanic Activity For Essential Services In Montserrat: Challenges, Adaptations And Resilience. In: Wadge G, Voight B, Robertson R (eds) The Eruption Of Soufrière Hills Volcano, Montserrat From 2000 To 2010, Geological Society, London.
52. Tanguy J-C (1994) The 1902–1905 eruptions of Montagne Pelée, Martinique: anatomy and retrospection. J Volcanol Geotherm Res 60(2):87-107
53. Tierney K (2012) Disaster governance: social, political and economic dimensions. Annu Rev Environ Resour 37:341-363
54. Twigg J (2004) Disaster Risk Reduction: Mitigation And Preparedness In Development And Emergency Planning. Overseas Development Institute, London.
55. (2010) Disaster management capacity assessment report: Montserrat. United Nations Development Programme in Barbados and the. OECS, Bridgetown.

56. Wadge G, Isaacs MC (1986) Volcanic hazard from Soufrière Hills Volcano Montserrat. A report to the Government of Montserrat and the Pan Caribbean Disaster Preparedness and Prevention Project. Department of Geography, University of Reading, Reading.
57. Whitten SM, Hertzler G, Strunz S (2012) How real options and ecological resilience thinking can assist in environmental risk management. J Risk Res 15(3):331-346
58. Wilkinson E (2012) Why "small is beautiful" in municipal disaster risk reduction: evidence from the Yucatán peninsula, Mexico. Environ Hazards 11(2):155-171
59. Wilkinson E (2013) Disaster Risk Governance. In: Volcanic Areas: A Concept Note For Work Package 4 Of The Strengthening Resilience In Volcanic Areas (STREVA) Programme. Overseas Development Institute, London.
60. (2002) Natural Hazard Risk Management. In: The Caribbean: Revisiting The Challenge, vol Report No. 24166-LAC, 1. Caribbean Group for Co-operation in Economic Development, World Bank, Washington D.C.

Chapter 4

Risk Perception in Fire Evacuation Behavior Revisited: Definitions, Related Concepts, and Empirical Evidence

Max T Kinateder[1], Erica D Kuligowski[2], Paul A Reneke[2], and Richard D Peacock[2]

[1]Department of Cognitive, Linguistic, and Psychological Sciences, Brown Unversity, Providence, RI, USA

[2]National Institute of Standards and Technology, Gaithersburg, MD, USA

ABSTRACT

Risk perception (RP) is studied in many research disciplines (e.g., safety engineering, psychology, and sociology). Definitions of RP can be broadly divided into expectancy-value and risk-as-feeling approaches. In the present review, RP is seen as the personalization of the risk related to a current event, such as an ongoing fire emergency; it is influenced by emotions and prone to cognitive biases. We differentiate RP from other related concepts (e.g., situation awareness) and introduce theoretical frameworks relevant to RP in fire evacuation (e.g., Protective Action Decision Model and Heuristic-Systematic approaches). Furthermore, we review studies on RP during evacuation with a focus on the World Trade Center evacuation on September 11, 2001 and present factors modulating RP as well as the relation between perceived risk and protective actions. We summarize the factors that influence perception risk and discuss the direction of these relationships (i.e., positive or negative influence, or inconsequential) and conclude with presenting limitations of this review and an outlook on future research.

INTRODUCTION

Occupants need to reach a place of safety during building fire emergencies. Evacuation behavior enables building occupants to do so (ISO/IEC [2008]). Figure 1 gives an overview of the evacuation process and it illustrates that occupant evacuation from buildings comprises two distinct periods: *pre-evacuation* and *evacuation period* (Kuligowski et al. [2010]). The pre-evacuation period can be further split into a pre-alarm phase, an information seeking phase, and a response phase (in which initial protective actions are taken); it ends when an evacuation decision is made (Purser and Bensilum [2001]). The crucial point in the pre-evacuation period is the decision of occupants to evacuate after they have received initial fire cues[a], which marks the transition from pre-evacuation to evacuation behavior. This decision is potentially dependent on occupants' risk perception (RP) and other human factors. For a recent review of human factors in building evacuation, see Ronchi and Nilsson ([2013]).

[a]In the present article, the term *fire cue* refers to all cues provided in a scenario initiated by a fire. These are not restricted to fire effluent cues and include indirect indicators of a fire emergency (e.g., seeing other occupants evacuating or receiving information via a public address system).

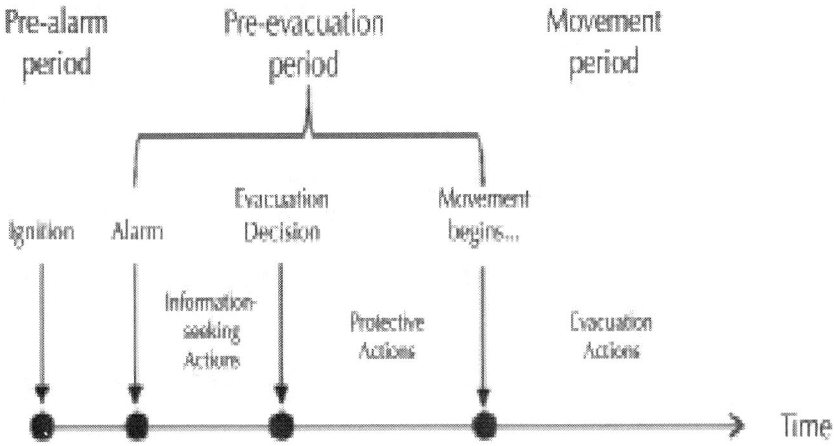

Figure 1: Timeline of building fire evacuation.

Engineering tools such as evacuation computer models, aim to establish the Available/Required Safe Egress Time (ASET/RSET) of a building. RSET is defined as the time occupants need from the fire onset until they reach a place of safety. ASET refers to the time which is actually available for evacuation, specifically the time between the onset of a fire and the point at which conditions become fatal (Kobes et al. [2010]). Most evacuation models implement oversimplified assumptions about the pre-evacuation period. For example, psychological processes and social interactions are often not considered. For an overview, see Kuligowski et al. ([2010]) or Proulx ([2008]). This increases the uncertainty in evacuation models, as studies have shown that the pre-evacuation period can be as long as or longer than the actual evacuation time period (Proulx [1995]; Fahy and Proulx [2001]; Kobes et al. [2010]).

It is important to understand RP during building fire evacuations for many reasons. First, since RP is hypothesized to influence the point of transition from pre-evacuation to evacuation (or other protective) behavior, it is questionable whether an accurate description of the evacuation process is possible in the absence of an accurate description of the RP. In the worst case, faulty assumptions about RP may find their way into evacuation models or affect building design. Second, understanding RP and its relevance for evacuation decision-making may in turn contribute to the development of more accurate evacuation models, via more precise predictions of ASET/RSET, and ultimately improve building safety. A significant part in this endeavor is the eventual development of a comprehensive behavioral theory on human behavior in fire (Kuligowski and Gwynne [2010]). A comprehensive theory of human behavior in fire would describe and explain aspects of evacuation behavior in logical terms that are consistent with systematic observations of the real world.

A review of the literature on the topic of RP has highlighted a variety of ways in which RP has been discussed. First, research studies on the topic often attempt to identify the factors that influence perceived risk. These factors can be individual-based (i.e., psychophysiological states or traits of an individual), physical (i.e., in the environment) or social (i.e., the behavior of others) in nature. Second, research studies have questioned whether RP influences aspects of the evacuation process such as the evacuation decision or evacuation delay time (i.e., the time occupants wait before they start evacuating). In either case, literature on RP and evacuation often does not propose a definition of RP or the way in which the research has defined the term (See Table 1 for an overview of different ways of operationalization of perceived risk in research studies). Therefore, the first goal of this literature review is to clarify the concept of RP in the context of building fire evacuation, and to provide a definition of RP specifically for the field of fire protection engineering. This includes the distinction from similar relevant concepts (e.g., situation awareness) and a discussion of the scope (e.g., the spatial and temporal proximity of a threat) of RP is presented.

Table 1: Overview of studies on RP and evacuation

Ref.	Rel.[1]	Scenario	N	Study population	Transfer to building fires possible?[2]	Qual./Quan.	Control group	Data
Kuligowski and Mileti (2009)	[illegible]	Building evacuation under a terrorist attack	[illegible]	WTC occupants[3]	yes	Quan.	no	Retrospective
Day et al. (2013)	[illegible]	Building evacuation under a terrorist attack	[illegible]	WTC occupants[4]	yes	Quan.	no	Retrospective
Kuligowski (2013)	[illegible]	Building evacuation under a terrorist attack	[illegible]	WTC occupants[4]	yes	Qual.	no	Retrospective
Sherman et al. (2011)	[illegible]	Building evacuation under a terrorist attack	[illegible]	WTC occupants	yes	Quan.	no	Retrospective
Gershon et al. (2007)	[illegible]	Building evacuation under a terrorist attack	[illegible]	WTC occupants	yes	Qual.	no	Retrospective
Gershon et al. (2012)	[illegible]	Building evacuation under a terrorist attack	[illegible]	WTC occupants	yes	Qual.	Comparison to WTC occupants who were not in the building at the incident	Retrospective
[illegible] et al. (2013)	3	Tunnel accident and fire	[illegible] tunnel fires	Tunnel users	With limitations	Qual.	no	Retrospective
Averill et al. (2005) Averill et al. (2009)	[illegible]	Building evacuation under a terrorist attack	100	WTC occupants[3]	Yes		no	Retrospective
McConnell et al. (2010)	2	Building evacuation under a terrorist attack	126	WTC occupants	yes	Quan.	no	Retrospective
[illegible] et al. (2013)	2, 3	Elevator evacuation during an unspecified emergency	[illegible]	High-rise building occupants	Yes, with limitations	Quan.	no	Cross-sectional
[illegible] and [illegible] (2012)	3	Accident in chemical/nuclear facility	[illegible]	Employees in chemical & nuclear facility	With limitations	Quan.	no	Cross-sectional
Baker et al. (1991)	3	Hurricane evacuation	[illegible]	Residents in hurricane risk regions	With limitations	Quan.	no	Retrospective
Brenkert-Smith et al. (2013)	3	Wildfire evacuation	[illegible]	Wildland urban interface (WUI) homeowners in Boulder and Larimer Counties in Colorado, USA	Yes, with limitations	Quan.	No	Prospective
Dash et al. (2007)	3	Hurricane evacuation	706–407	General population in hurricane area	With limitations	Quan.	no	Retrospective
Matyas et al. (2011)	3	Hurricane evacuation	148	Tourists	With limitations	Quan.	no	Cross-sectional
Horney et al. (2010)	3	Hurricane evacuation	570	General public	With limitations	Quan.	no	Retrospective
Mozumder et al. (2008)	3	Wildfire evacuation	[illegible]	Fulltime & seasonal residents	With limitations	Quan.	no	Retrospective
Siebeneck and Cova (2012)	3	Flood evacuation	[illegible]	General population in flood area	With limitations	Quan.	no	Retrospective
Drabek (2001)	3	Natural disaster	406	Business employees	With limitations	Qual.	no	Retrospective

Kuligowski and Mileti (2009)		1 item (yes/no): "During the time when you first became aware that something had happened and when you first entered the stairwell or elevator to leave did you believe that other people were in danger of being killed?"	PADM	Environmental cues, floor level, obtained information,	No direct effect on evacuation delay (beta ≈ 0 for both towers); weak effect on information seeking behavior (beta ≈ 0.11); in one tower, pre-evacuation actions were associated with higher perceived risk (beta ≈ 0.23 vs. beta ≈ 0.08)
Day et al. (2013)	Interview	[illegible]-point Likert scale ("How much at risk did you feel?")	-	Number of cues, quality of cues, distance to impact	High perceived risk predicted early responders (Beta = .36, OR = 1.44); Low perceived risk was not a predictor of delayed evacuation
Kuligowski (2013)	Interview	[illegible]-point Likert scale	PADM	Previous experience, hyper vigilance, cue intensity, cue identification	Perceived risk predicted evacuation decision
Sherman et al. (2011)	Questionnaire	1 item asking "How serious did you think the situation was at first?" on a [illegible]-point Likert scale		female, member of port authority NY/NJ (personal background variables), evacuating from WTC 1 (vs WTC 2), more Environmental Cues, more unusual Event (context variables), lower education, longer tenure in the towers, more knowledge, more emergency preparedness	lower perceived risk: - less information seeking - more pre-evacuation actions - longer pre-evacuation delays (beta = -.25)
Gershon et al. (2007)	In-depth interviews (n = [illegible]) or focus groups (n = [illegible])	Coding of qualitative interviews	-	-	Yes, emergent perception of risk formed by sensory cues facilitated evacuation decisions (but not the process of evacuation)
Gershon et al. (2012)	Questionnaire	Several items (number not specified) including seriousness of the situation, and concerns that the building would collapse	Behavioral Diagnostic Model		Yes, 70% stated that they evacuated because they appraised the situation as dangerous. Occupants who thought the situation was serious evacuated with less delay (OR = [illegible]) and faster (OR = 1.80)
[illegible] et al. (2013)	Review of reports, video footage, media reports	Not reported	Danger control model	Visibility of exit	Yes
Averill et al. (2005) Averill et al. (2009)	Interviews			Seek info, environmental cues	
McConnell et al. (2010)	Questionnaire	[illegible]-point Likert scale	-	Floor level in tower, WTC1, time (before or during evacuation)	-
[illegible] et al. (2013)	Hypothetical scenario questionnaire	Rating of perceived safety of evacuation modes (two [illegible]-point Likert scale items)		Building floor, evacuation method (elevator vs. staircase)	yes

Mbaye and Kouabenan (2013)	Questionnaire		-	locus of control, positivity bias, availability heuristic	-
Riad et al. (1999)	Interview		-	-	yes
Brenkert-Smith et al. (2013)	Questionnaire	2 questions on perceived probability scaled to range from 0 to 100 and Likert scale for 4 variables on perceived consequences	Social amplification of risk framework	Lot size, Previous experience, social context	-
Lindell et al. (2005)	Questionnaire	-	-	-	-
Matyas et al. (2011)	Questionnaire	5 point Likert scale	-	-	Yes (correlated with stated preference)
Horney et al. (2010)	Interview	3 point scale (low-middle-high)	PADM	Actual risk, homeownership.	no
Martin et al. (2009)	Questionnaire	5 point Likert scale	PADM	Fire experience, subjective knowledge, perceived responsibility	Yes, mediated, 38% of variance in perceived risk explained
Siebeneck and Cova (2012)	Questionnaire	5 point Likert scale	Threshold model of RP	Distance to threat, Time course of events, amount of property damage	Not reported
Drabek (2001)	Questionnaire	4 items measuring risk-related behavior and perceived safety	Stress-strain perspective	Higher perceived risk was associated with lower amount of community disaster planning, warning messages implying that evacuation was mandatory, residing in a mobile home or apartment, working in a more formalized company, working in a younger company, and long-term event or consequences	Perceived risk predicted evacuation delay (beta = .145) multiple evacuation (beta = .158)

Note: The content of this table is solely based on the information available in the individual studies and the amount and accuracy of the reported information varies. Ref. = Reference; Rel. = Relevance; N = sample size; Quan. = Quantitative study; Qual. = Qualitative study; WTC = World Trade Center; [1]NIST WTC evacuation data base; [2]yes, with limitations, no, unclear; [3]If yes, describe the relation (e.g. mediated, correlated); [3] 1 = planned evacuation from a latent threat, 2 = acute evacuation from an acute threat than building fire, 3 = Fire evacuation from buildings; [4]labeled as milling in this study; [5]HEED data base; [6]no specification of actual number of participants was given in this paper.

The studies are sorted according to their relevance for RP and evacuation.

Kinateder et al.

Kinateder et al. Fire Science Reviews 2015 4:1 doi:10.1186/s40038-014-0005-z

When studies on RP are presented, researchers have often identified some theoretical underpinning of risk perception that provides the foundation for the methods in the study. Thus, the second goal of this review is to identify and describe relevant theoretical frameworks of RP from evacuation research and other disciplines.

The third goal is to present a systematic overview, summary and discussion of the factors affecting RP during fire evacuation. Specifically, the current knowledge on the role of RP in the pre-evacuation and evacuation period in the fire evacuation process as well as factors modulating the relation between RP and protective actions are discussed. This way, the present overview may contribute to theory development in fire evacuation research.

METHODS

For the purpose of the present literature review, we followed the steps for a systematic literature review suggested by Khan et al. ([2003]):

Step 1 Framing Questions for A Review: The following main research questions were formulated: What is RP? And what role does RP play during building fire evacuation? These questions comprise the headings for the main chapters of this review. Each of these two very broad questions was subdivided into several steps which represent the sub headings in the each chapter.

Step 2 Identifying Relevant Work: Relevant literature on RP was primarily identified by searching literature data-bases (Web of Science, Google Scholar, Science Direct, Social Science Research Network, EvacMod.net). The keywords used to identify relevant literature included the following terms: *risk perception, evacuation, fire emergencies, human factors, human behavior in fire, hazard perception, egress, disaster, situation awareness, threat awareness, risk assessment, perceived vulnerability, arousal, risk communication, safety climate, safety culture, hurricane evacuation, heuristics, systematic information processing,* and *decision-making.* The sources were accessed through the libraries of the National Institute of Standards and Technology and the University of Würzburg, Germany. For literature without full text access from either of these two libraries or through interlibrary loan, abstracts were considered, or the source was ignored. The literature identified included but was not limited to reports and journal articles from fire research, psychology, sociology, and biology. The search results were integrated with relevant literature from colleagues and other publications.

Step 3 Assessing the Quality of Studies: Literature was included if it was relevant to the topic. Only publications in peer reviewed journal articles, conference proceedings, or books from established scientific publishers were considered. The literature research was not restricted to a certain time period, journal, field, or geographical location. An important criterion was the precision of the description of study protocol, sample, data collection, and analysis methods. Since studies from a variety of fields were included at this point, studies were ranked according to their relevance to RP and fire evacuation (Table 1).

Step 4 Summarizing the Evidence: The main findings of the first question (What is RP?) are summarized in text in Section What is RP? Defining RP during fire evacuation. The results for the second question (What is the role of RP during fire evacuation?) are summarized in text and tables in Section What role does perceived risk play in building fire evacuation? and Section Overview of studies. The summaries address differences regarding the theoretical foundation, methods of data collection and analysis, as well as the interpretation of results of individual studies.

Step 5 Interpreting the Findings: The methods, results, and their implications are discussed, followed by a discussion of the limitations of the present literature review (Section Limitations). Finally, future research questions for the topic of RP in the field of fire safety engineering are identified (Section Conclusions and outlook).

WHAT IS RP? DEFINING RP DURING FIRE EVACUATION

Definition of RP during Fire Evacuation

In the context of fire evacuation, RP refers to the perception of an imminent threat to one's own life and health. Here, RP is defined as a psychological process that describes the subjective (conscious and unconscious) evaluation (as opposed to objective risk assessment) of the probability to be affected by an imminent undesirable event in a specific situation and an assessment of one's own perceived vulnerability and coping resources. RP is seen as the process of personalizing the risk related to the current event, such as an ongoing fire emergency. It is influenced by emotions and prone to cognitive biases. Note that the term RP refers to a psychological process with *perceived risk* as its outcome. RP can be differentiated from several similar and overlapping concepts, such as situation awareness, perceived vulnerability, hazard perception, risk assessment and threat awareness (see Section Related concepts and expressions). Theoretical frameworks (Section Theoretical frameworks on RP and evacuation) provide an understanding on how risk relevant information processing and coping mechanisms translate into behavior in general.

Other scientific definitions of RP in a wider sense refer to the subjective assessment of the probability of an undesired event, the magnitude of its consequences, and one's own coping capabilities (Michalsen [2003]; Rayner and Cantor [1987]; Wachinger et al. [2012]). In this context, coping capabilities refer to general and situation specific competencies of an individual (e.g., the ability to stay calm in stressful situations or expertise in firefighting). There are two main approaches to RP. The first can be summarized as an *expectancy-value approach* (Patterson et al.[2010]; Sjöberg et al. [2004]) and the second can be referred to as the *risk-as-feelings* approach (Slovic [2010a]).

According to the *expectancy-value approach*, RP consists of two components: an individual's assessment of a natural hazard and his/her perceived vulnerability (Patterson et al. [2010]). It comprises the beliefs (whether rational or irrational) held by an individual, group, or society about the likelihood, extent, magnitude, and timing of a threat; it refers to subjective assessments of probabilities of a specified type of accident happening, and how concerned one is with the consequences (Sjöberg et al. [2004]). Here, RP is seen as a conscious cognitive process which is prone to biases. In the case of building fires, this would reflect an evaluation to the self-posed question "Am I at risk?" after having received fire cues (e.g., a fire alarm or smoke).

The *risk-as-feelings* hypothesis criticizes the assumption that RP is an (entirely) conscious cognitive process (Loewenstein et al. [2001]; Slovic et al. [2005]; Slovic [2010a]). It stresses the role emotions play the moment decisions are made and it assumes that information needs to convey emotions in order to become meaningful for an individual. Here, RP refers to how much danger a person feels he/she is in as a result of the event (Sherman et al. [2011]). For building fires, this would reflect an occupant's "gut feeling" after perceiving fire cues.

Note that RP is seen as a subjective process of an individual. That is, RP is not necessarily related to objective risk (although potentially correlated) and is prone to various biases. One may hypothesize that both approaches are relevant and even connected for fire evacuation and simply refer to different aspects of how a building fire is experienced. Consequently, a holistic approach to RP in fire evacuation should include the expectancy-value as well as the risk-as-feelings approach.

The main difference between risk-as-feelings and the expectancy-value approach lies in the psychological processes, which may even be operating simultaneously (e.g., while walking through a dark empty street, one may feel at risk although one knows that one is in a safe area). Whereas the risk-as-feeling refers more to associative and emotional processes ("gut feeling"), the expectancy value process focuses on rational, cognitive processes. This differentiation is important for fire evacuation, since the results of the risk estimates of these processes can be different and consequently, behavior may vary depending on which approach is predominant.

Characteristics of RP

As RP is studied in many research disciplines (e.g., fire protection engineering, psychology, and sociology) (Wachinger et al. [2012]; Slovic [2000]), the contexts to which concepts of RP are applied vary greatly. The following paragraphs elaborate on the characteristics of RP and links them to the specific case of fire evacuation.

As the term suggests, RP comprises a *risk* and a *perception* component. 'Risk' has various meanings in everyday usage, such as hazard (e.g., What are the most important risks for occupants during a building fire?), consequence (e.g., What is the risk of delayed evacuation during building fires?), probability (e.g., What is the risk of being in a building fire?), or potential adversity or threat (e.g., What is the risk of being exposed to a building fire?) (Slovic and Weber[2002]). This highlights a critical aspect in many questionnaire studies on evacuation and RP in which participants were simply asked, 'how much risk' they felt (e.g., Day et al. [2013]; Martin et al. [2009]; Siebeneck and Cova [2012]; Matyas et al. [2011]; McConnell et al. [2010]; Horney et al. [2010]). It is possible that participants had different concepts about the term 'risk' when they rated their perceived risk. Note that these lay concepts of risk vary significantly from the scientific definition in fire safety, where risk is "the potential for realization of unwanted, adverse consequences to human life, health, property, or the environment. (Watts and Hall [2008], p. 3)"

'Perception' is defined as the organization, identification, and interpretation of sensory information in order to represent and understand the environment (Schacter et al. [2011]). RP bridges all perceptive modalities and comprises various cognitive processes (e.g., sense-making, decision-making, or appraisal). In this context, perception can be understood as a signal-detection process, where occupants continuously scan their environment with their senses and have to filter threat-relevant fire cues from the noise of irrelevant input. This process results either in a hit (correct detection), miss (cue not detected/ incorrect rejection), false alarm, or ignore (correct rejection). The criterion as well as the signal-to-noise ratio affects the signal detection (See Green and Swets ([1966]) for an introduction to signal detection theory). Several factors, such as previous experience with fire, may lower the threshold criterion of detection or increase the sensitivity to fire cues (leading to more hits and false alarms). The more complex an environment becomes, the more the amount of sense-based "noise" increases, which makes it more difficult for an individual to differentiate fire cues from irrelevant stimuli (more misses). In other words, the fire cues become less salient for the occupants.

Scope of RP Research

The scope of RP research varies across disciplines and it is questionable if and how results can be transferred from one field to the other. In fact, RP studies on technological threats and natural hazards vary significantly in their outcome (Wachinger et al. [2012]). Specifically, the scope of RP research can be described in a framework of threat certainty, the time frame of risks, and the target of RP.

Threat certainty refers to how likely an undesired event is and threats can be categorized into imminent or latent threats. Imminent threats are certain to occur, very near, or impending and require immediate responses. A latent threat refers to threats from potential disasters associated with certain risk factors, such as living in a high-risk hurricane or earthquake region. Here, the incidence of the actual event is not predictable (for an individual) in the foreseeable future. Most of the literature on RP during disasters covers latent threats in which consequences are uncertain, rare, and/or delayed. For emergency evacuation during fire, however, imminent threats are relevant.

The *time frame of risks* can be differentiated into short term and long term risks. Long term risks refer to risks that lie relatively far in the future (e.g., hurricanes that are near the coast for days in advance); for the present paper, short term risks lie in the immediate future (within hours, minutes or even less time). The long term perspective could be seen as the general tendency of a person to expect a threat. In the case of building fire evacuation, the time frame of risks is most likely short term.

The *target of RP* addresses the question of what is at risk for an individual. RP can be directed at various aspects of life; including one's life and well-being, status, property, goals, and also others. For fire evacuation, the RP of one's own life and health seems most relevant. This is also sometimes referred to as *personal risk* (Perry [1979]). However, it is possible that other aspects of RP may compete with one's own safety (Firing et al. [2009]). For example, family members or significant others in a residential evacuation may act as a modulating factor for one's own personal risk.

Related Concepts and Expressions

The following section describes several concepts that either overlap with the present definition of RP or are sometimes used synonymously. Of these, *situation awareness* is the most relevant in the context of fire evacuation and will be discussed in more detail. Given the conceptual overlap, for example, in the importance of appraisal processes in RP and situation awareness, these related terms were also considered during literature research.

Situation Awareness (or sometimes known as situational awareness) is a key concept introduced by Endsley in the decision-making literature (Endsley and Jones [2013]). Situation awareness is defined as "the perception of the elements in the environment within a volume of time and space, the comprehension of their meaning and the projection of their status in the near future" (Endsley[1988], p.97). It is conceptualized as an internalized temporal and spatial representation or mental model of a person operating in a complex environment (Endsley and Jones [2013]; Endsley[1988]; Sarter and Woods [1991]). The quality and precision of such mental models affect decision-making and depend, among others, on the complexity of the environment. Poor situation awareness has been identified as a major cause in accidents related to human errors (Endsley[1995]). Apart from the definition reported here, several other definitions can be found in the literature, of which a discussion is beyond the scope of this paper (See Sarter and Woods ([1991]) for a detailed summary of situation awareness concepts). Although closely related, situation awareness is a more holistic concept than RP as it applies to the environment as a whole and not only to hazards. Furthermore, situation awareness can be seen as a conscious process, whereas RP consists of conscious (expectancy-value assessments) and unconscious components (the feeling of risk). Some authors argue that RP can be understood as situation awareness for dangerous situations (Horswill and McKenna [2004]). Although RP and situation awareness overlap, they are independent concepts. Individuals may feel at risk with both high and low situation awareness.

Perceived vulnerability is the subjective appraisal of one's own capacity to anticipate, cope with, resist and recover from the impact of a natural hazard (Wisner [2004]). Some authors use RP and perceived vulnerability synonymously (Riad et al. [1999]).

Hazard perception is the skill to detect developing threats (McKenna and Crick [1991]). This concept is mainly used in traffic research. Some authors argue that hazard perception reflects situation awareness for dangerous situations in the traffic environment and improves with training (Horswill and McKenna [2004]).

Threat awareness (related *death awareness*) can be understood as the general mental model an individual has about life threatening events (Hirschberger et al. [2009]). It is part of terror management theory, which addresses how humans cope with the idea of their own mortality (Greenberg et al. [1990]).

Risk assessment is the "identification, evaluation, and estimation of the levels of risks involved in a situation, their comparison against benchmarks or standards, and determination of an acceptable level of risk" (Business Dictionary [2013]). Most of the literature using this term refers to *objective*risk assessment as compared to RP which is subjective. Objective risks can be statistically estimated (e.g., the calculation of probability and estimated damages of environmental disasters). Similar to RP, the time frame, scope, and certainty of a threat can vary in risk assessment. Here, risk is conceptualized as the product of probability and consequences of an undesired event (Yuan et al. [2009]).

Risk communication is the field of research that deals with the exchange of information and education about risk-related content. Risk communication is relevant to a wide range of contexts (e.g., avoiding industrial accidents, illnesses, traffic, disasters) (Wogalter et al. [1999]; Fischhoff[1995]). Its importance lies in the fact that successful risk communication can lead to improved safety behavior without having to learn from experience. For the case of fire evacuation, risk communication may contribute to an increased awareness and preparedness of occupants as well as to more effective evacuation behavior. Risk communication affects RP.

Safety climate refers to a (work or living) community's shared perception of their organization's policies, procedures, and practices as they relate to the value and importance of safety within the organization (Griffin and Neal [2000]). Safety climate may affect RP and other factors such as situation awareness.

Safety culture summarizes the shared values and beliefs in an organization which interact with its structures and control systems to produce safety related behavioral norms (Thompson et al.[1996]). Similar to safety climate, safety culture influences RP.

Arousal refers to the general activation of the sympathetic nervous system. Although not directly related to RP, arousal may be strongly correlated with the underlying physiological and psychological processes of RP. For example, arousal affects decision-making, in the sense that higher arousal is related to more impulsive decision-making (Strack and Deutsch [2004]).

Fear is an emotional response to a perceived threat and a common reaction to emergency situations (Öhman [2000]). Animal studies sometimes use fear reactions as an indicator of RP (Stankowich and Blumstein [2005]).

Theoretical Frameworks on RP and Evacuation

Since RP is relevant to multiple disciplines, several theoretical frameworks addressing RP have been developed. Nonetheless, very few studies on RP during evacuation mention being founded on a specific theory or theoretical framework (See Table 1). The following sections introduce theoretical models related to RP and human behavior in emergency situations. Most of the theories follow the *psychometric paradigm*, which is the basis of the risk-as-feeling approach (Slovic[2000]) and aims to develop objective, reliable and valid measurement tools of psychological processes (e.g., rating scales or standardized questionnaires) (Eignor [2013]). Within the psychometric approach, quantitative subjective ratings of perceived and acceptable risk are mapped together with the desired level of control for a given situation and compared to ratings of other situations (Slovic [1987]).

Heuristic-Systematic Models

The *Heuristic-Systematic* approach refers to two-process models of information processing and can be applied to RP. Such models are widespread in the psychology literature (e.g., Strack and Deutsch [2004]; Chaiken and Maheswaran [1994]; Chaiken and Eagly [1989]; Chaiken [1980]; Kahneman [2011]). The basic assumption is that information can be processed systematically, heuristically, or in a combination of the two. In systematic information processing, all available information is assessed according to its meaning and relevance. This process is thorough but also demands more time and resources. *Prospect theory* (Kahneman and Tversky [1979]) first introduced the concept of heuristics, which can be understood as mental shortcuts or simple rules of thumb that allow for the making of fast decisions at the cost of less systematic information processing. Heuristics are useful tools for decision-making if sufficient information about probabilities or time and other resources for a slower and more thorough assessment are not available. For example, research on natural disasters has shown that the actual probability of an event is rarely regarded in risk appraisals (Miceli et al. [2008]). Several types of heuristics are relevant for evacuation and RP:

The *affect heuristic* refers to the fact that current emotional states influence decision-making. This concept is an important part of the risk-as-feeling approach. Emotional states modulate the understanding of numbers and probabilities. For example, large numbers are underweighted in decisions and lack meaning for people unless they convey a feeling (Slovic [2010b]). Similarly, judgments of risk and utility are often influenced by whether one likes something, e.g., utility is overestimated and risk underestimated for activities associated with positive emotions (Slovic et al. [2005], [2007]; Finucane et al. [2000]).

Anchor Heuristics describe the tendency to overly rely on a few initial or salient pieces of information (anchors) when making a decision. Subsequent or less salient cues may be ignored or weighed less. Anchoring itself can be affected by mood, expertise, or the use of other heuristics (Furnham and Boo [2011]). This may lead to over or underestimation of risk during fire evacuation. For example, an occupant might interpret the sound of a fire alarm as a cue for a drill and subsequently ignore more subtle cues of a real incident.

Availability heuristics describe how likelihood estimates of an event are affected by how easy it is to recall or imagine it (Kahneman and Tversky [1979]). The more "available" an event is in memory, the higher its estimated likelihood (Greening et al. [1996]). For example, occupants may assess the risk of a fire emergency by the ease of recalling similar occurrences.

Representativeness heuristic notes that likelihood estimates of an event are often judged by their similarity to its parent population (Kahneman and Tversky [1979]). The more a cue seems to fit into a certain category of events, the more likely it will be estimated as indicative of it. For example, occupants may perceive a fire alarm sound as not indicative for a fire emergency if it sounds similar to other alarm sounds (e.g., an error sound from an electronic device).

Similarly, *proximity heuristics* describe the "tendency to judge probabilities by monitoring the spatial, temporal, or conceptual distance to a target" (Teigen [2005], p. 424) and has been studied to understand estimates of accident probabilities. An example of a proximity heuristic might be the case of occupants overestimating the probability or severity of a fire emergency if they perceive fire cues matching their expectations about a fire emergency scenario.

The use of heuristics may explain another type of bias known as *normalcy bias*, which refers to the tendency to interpret cues as indicative for everyday events and underestimating the likelihood and consequences of disasters (Okabe and Mikami [1982]). During building fires, when occupants are faced with ambiguous information, the normalcy bias is likely to last longer while occupants remain inside the building (Kuligowski and Gwynne [2010]). Additionally, the anchor, availability and representativeness heuristics may lead to low perceived risks, since many fire cues are not specific to an emergency. For most cases, the assumption that a fire alarm is just another drill and not a real emergency is true. During the evacuation of the World Trade Center (WTC) on September 11, 2001, occupants, especially those from the lower floors, reported relatively low RP which may be attributed to the assumption that nothing extraordinary was going on in the building (McConnell et al. [2010]).

When processing information *systematically*, individuals aim to understand the available information and its relevance for RP and decision-making. This process is relatively slow and requires significant resources. In heuristic information processing, RP is based on relatively automatic processes in which little effort is spent on processing the information (Smerecnik et al.[2012]).

Whether information is processed systematically or heuristically depends on an individual's level of arousal (i.e., activation of the sympathetic nervous system), available cognitive resources, and other factors, such as experience, emotional states, or personality traits (Strack and Deutsch[2004]). Perceived risk, as defined above, may also determine whether information is processed heuristically or systematically. In line with this, a study by Sherman et al. ([2011]) on building evacuation suggested a curvilinear relationship between perceived risk and information seeking behavior, an indicator of systematic processing. Participants reporting either low or high perceived risk were less likely to seek more information as opposed to those with a medium level of perceived risk (Sherman et al. [2011]). Both systematic and heuristic processes can lead to an evacuation decision, but they may be affected by different factors. Both processes are prone to biases and limited within each individual.

Similar to the systematic and heuristic approach, the concept of *bounded rationality* describes that decision-making is limited by the available information, the cognitive resources, and the finite amount of time to make a decision. Here, the term *satisficing*, describes the process in which occupants do not base their decisions on all available information but on the amount of information they deem sufficient for their decision (Simon [1972], [1991]). Satisficing could be understood as a form of heuristic information processing. Drabek developed the *stress–strain perspective* based on the concept of bounded rationality (Drabek [2001]). In line with the heuristic-systematic approach, constraints (e.g., the availability of information) in the social and physical environment are hypothesized to bias RP and behavior during emergencies.

Appraisal Models

Several psychological models underline the importance of appraisal processes for decision-making. The most prominent of these models is the *Transactional Stress Model*, which is not a RP model per se but provides insights on coping mechanisms when people are faced with risk (Lazarus and Folkman [1984]). It is a classic cognitive theory on emotion regulation, is closely linked to RP, and postulates several appraisal processes.

Primary appraisal: "How relevant is this situation to my needs?" Is there the risk of harm, loss, threat, or challenge? In the case of evacuation, this is the assumed reaction to the alarm signal. If the alarm is deemed relevant, the next appraisal process follows.

Secondary appraisal: "Do I have the necessary resources available to cope with the situation?" If yes, then problem-focused attempts to cope with the situation are used. If no, then emotion-focused coping is used (i.e., If I cannot change the situation, I have to adapt my emotional reaction to it).

Re-appraisal after coping attempts: "How is the situation now?"

Problem-focused coping does not automatically imply that occupants would choose adequate reactions. Classic cognitive stress models, such as the transactional stress model, focus on the subjectively perceived threat of a situation (Lazarus and Folkman [1984]), which can be interpreted as RP. Psychological stress occurs if one does not possess the necessary resources to cope with a situation which is perceived as dangerous.

Appraisal processes, similar to the ones discussed in the *Transactional Stress Model*, have been incorporated into theories developed specifically for human behavior in fire. Proulx's cognitive stress model of people facing fire, for example, takes into account different factors, such as information processing, decision-making, problem-solving, and stress (Proulx [1993]). Similar to the Transactional Stress Model, Proulx sees iterative appraisals of the situation and one's own coping resources at the core of experienced stress and behavior. According to this model, several*stress loops* are triggered when occupants are confronted with a fire outbreak, in which the appraisal of ambiguous information and increased danger can lead to fear, worry, and confusion (Proulx [1993]).

The importance of appraisal processes during catastrophic events has been empirically studied. For example, in a questionnaire study with hurricane survivors, Riad, Norris, and Ruback found that 58% of the respondents chose not to evacuate from a severe hurricane threat. The most important reasons for not evacuating during a hurricane were that the hurricane had not been perceived as a serious threat (primary appraisal), participants had been confident that their current place was as safe as any other (secondary appraisal), and participants avoided thinking about the situation (coping, Riad et al. [1999]). The misinterpretation of valid threat indicators may therefore be a key problem in the process of evacuation. Evidence from a hypothetical scenario study showed that participants appraised different types of disasters (crime, natural disaster, terrorist attack) as similar in risk, but they differed in the intentions to take protective actions. For example, in a natural disaster scenario participants were more likely to state they would change their daily activities than in a crime scenario (Heilbrun et al. [2010]). The cognitive appraisal of a given situation as dangerous may influence the readiness to engage and the choice of protective actions. For example, a recent meta-analysis showed that the motivation to participate in safety trainings rises if the consequences of a potential event are perceived as threatening (Burke et al.[2011a]).

Protective Action Decision Model

The *Protective Action Decision Model* (PADM) was developed to provide a holistic approach to human behavior in emergency situations. It sets up a descriptive framework of the information flow and decision-making that affects protective actions taken in response to disasters (Lindell and Whitney [2000]; Lindell and Perry [1992]; Kang et al. [2007]; Huang et al. [2012]; Houts et al.[1984]; Kuligowski [2011]). The model describes the path from the initial perception of hazard cues to the initiation of protective action. It takes a variety of predispositions, such as environmental or social context, into account. Furthermore, it stresses the importance of appraisal processes, and thus links cognitive psychological approaches, such as the aforementioned transactional stress model, with classic safety engineering models.

A brief overview of the processes in the PADM follows with a discussion of the role of RP in the model. For a more comprehensive description of the model, see Kuligowski ([2011]); (Kuligowski[2013]).

PADM differentiates between pre-decisional and decisional processes. The former are the basis on which an individual makes his/her evacuation decision. The pre-decisional processes comprise (1) perceiving, (2) directing attention to, and (3) comprehending relevant fire cues. If the three pre-decisional processes have identified potentially relevant fire cues, occupants are hypothesized to engage in a five step decision-making process which may result in protective actions (Lindell and Perry [2012]; Perry and Lindell [2004]):

- Risk identification: Is there a real threat that I need to pay attention to? [If yes, then the occupant believes the threat]
- Risk assessment: Do I need to take protective action? [If yes, then the occupant decides that he/she needs to take protective action]
- Protective action search: What can be done to achieve protection? [The occupant begins searching for possible protective action strategies]
- Protective action assessment: What is the best method of protection? [The occupant chooses one of the action strategies developed in the previous stage and develops a protective action strategy or plan]
- Protective action implementation: Does protective action need to be taken now? [If yes, the occupant follows the plan developed in the previous stage]

As stated earlier, RP is defined in this review as the subjective evaluation of the probability to be affected by an imminent undesirable event and the assessment of one's own perceived vulnerability. This corresponds to the two first decisional processes in PADM. Lindell and Perry incorporate threat perception into PADM, which can be understood as an equivalent to primary appraisal in the transactional stress model (Lindell and Perry [2012]; Lazarus and Folkman[1984]). Their approach to RP corresponds to the expectancy-value approaches discussed earlier and also includes emotional and motivational aspects (labeled dread and unknown risks) (Lindell and Perry [2012]). The first stage of the decision model involves hazard identification in which the properties of a potential threat have to be evaluated. In the second step, risk assessment, ones' own vulnerability toward the threat is judged. This clearly represents the cognitive side of RP (systematic assessment of expectancy and values). One may speculate that the *risk-as-feeling*side is at least implicitly part of the pre-decisional as well as the risk identification and assessment processes.

It is also possible to integrate heuristic processing into PADM. Heuristics could play two different roles in the PADM. First, heuristics and systematic processing may be competing at each decisional step. Depending on the arousal, cognitive resources, previous experience, or the result of one of the decisional processes, an occupant may process each of the five decisional questions heuristically or systematically. For example, an occupant who identified a potential risk and sees him or herself as extremely vulnerable may rely on heuristics to identify protective actions. Second, heuristics may lead to skipping some of the decisional processes and thus speed up the decision-making at the cost of less thorough reasoning. Consider, for example, the anchor heuristic. An occupant might interpret the sound of a fire alarm as a cue for a fire emergency and immediately start evacuating without going through the steps of protective action search and assessment.

It is important to understand how RP affects evacuation activities. As theorized by the PADM, RP can be understood as a threshold mechanism for evacuation decision-making (Siebeneck and Cova [2012]; Kates [1971]). Therefore, it is possible to hypothesize that there is a threshold of acceptable risk for an occupant before he/she decides to evacuate. Evacuation decision-making is "triggered" if the perceived risk becomes unacceptable.

The PADM is a descriptive model of decision-making during emergency situations. As such, it does not make predictions about future behavior. However, it is possible to integrate the processes described in the PADM into predictive models, such as the *Evacuation decision model (EDM)*. EDM aims to predict the point in time when the decision to take protective action is made and assumes that RP is the key factor in this process (Reneke [2013]).

Reasoned Actions Models

Reasoned actions models, such as the Theory of Planned Behavior (TPB, Figure 2) or the Theory of Reasoned Action (TRA), are general theories describing how intentions are transferred into actions (Ajzen [2011]; Sheppard et al. [1988]). They fall into the systematic branch of the heuristic-systematic approach. These models assume that "intentions are the immediate antecedents of behavior and intentions themselves are a function of attitude toward the behavior, subjective norm, and perceived behavioral control" (Ajzen [2011]). RP plays a role in an individual's assessment of his/her perceived behavioral control (i.e., Do I have the resources to change the odds for an undesired event?). Most applications of these models have been used to predict long term behavior (e.g., changes in health behavior). However, it seems possible to apply the TPB to planned evacuation behavior. One study applied the TRA to hurricane evacuation behavior (Kang et al. [2007]). TRA assumes that occupants' conscious intentions to engage in a behavior are the principal determinants of actual behavior. The main limitation of this approach is that it is purely cognitive and leaves out affective situational variables (e.g. fear and anxiety). However, unanticipated barriers can arise between the intention and the opportunity to act, thus making the actual behavior different from the behavioral intention (Fishbein [1979]).

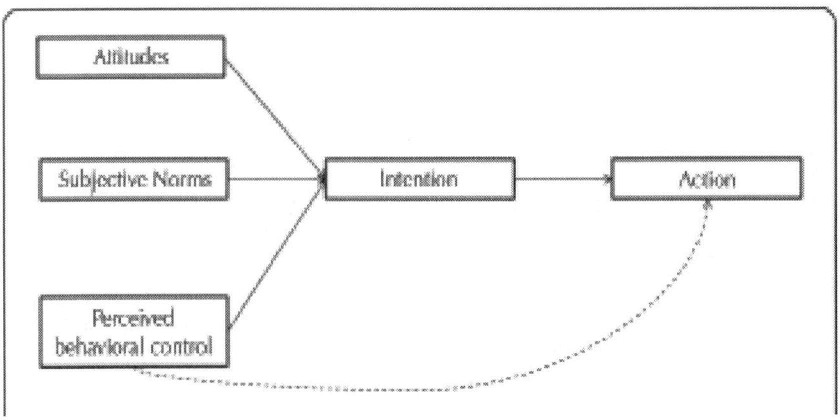

Figure 2: The Theory of planned behavior (TPB); Redrawn from Rogers & Prentice-Dunn ([1997]).

A meta-analysis of protection motivation models showed that increases in threat severity, threat vulnerability, response efficacy, and self-efficacy facilitated adaptive intentions or behaviors. Decreases in maladaptive response rewards and adaptive response costs also increased adaptive intentions or behaviors (Floyd et al. [2000]).

The *Protection Motivation Theory* (Rogers and Prentice-Dunn [1997]) is a model developed to understand and predict long-term health behavior. It tries to explain the effects of threatening (health) information on changes in attitude and behavior (e.g., planning to quit smoking after learning about the smoking related diseases). Protection motivation theory can also be applied to (planned) evacuation behavior. It hypothesizes that perception of the severity, susceptibility, or probability of occurrence, perceived self-efficacy, and perceived response efficacy modulate protective actions (Cauberghe et al. [2009]).

Although the reasoned action models were not developed to understand building fire evacuation, they have important similarities with other models (e.g., PADM) and may help to better understand RP and evacuation behavior. Unlike the more disaster specific models, the reasoned action models have been studied and found applicable to a wide range of behaviors. Thus we speculate that at least for planned evacuation, similar processes like the ones described in TRA or TPB can be assumed. However, these models do not apply to spontaneous and unplanned behavior and mostly ignore situational and affective variables. The important question is whether building fire evacuation is planned behavior or not. The answer to this question has consequences on how RP has to be conceptualized. If evacuation was a predominantly planned behavior, RP would most likely have to be understood from an expectancy-value perspective. If evacuation behavior does not involve long term planning, the risk-as-feeling approach may be more relevant. For most occupants, evacuation is clearly not a long term planned behavior in the sense that occupants plan the following: "When the fire alarm goes off, I will (not) evacuate". However, the time from the initial cue to the evacuation decision can be seen as a planning phase (as it is in the PADM) in which occupants appraise their situation, vulnerability, resources and options. Future studies should investigate to what degree evacuation from an imminent threat is planned behavior.

Hazard to Action Chain Model

Wachinger et al. ([2012]) developed the *Hazard to action chain model* (Figure 3) based on a literature review describing the effect of RP on protective actions during natural disasters. The authors assume that, similar to reasoned action models, intentions (labeled as "willingness to act") and preparedness are the precursors of (protective) actions. According to this model, RP influences preparedness as well as intentions: The higher the perceived risk, the higher the preparedness and intentions. The authors found that the most robust predictors of RP were trust in authorities (lower trust leading to higher perceived risk) and previous personal experience of a natural disaster.

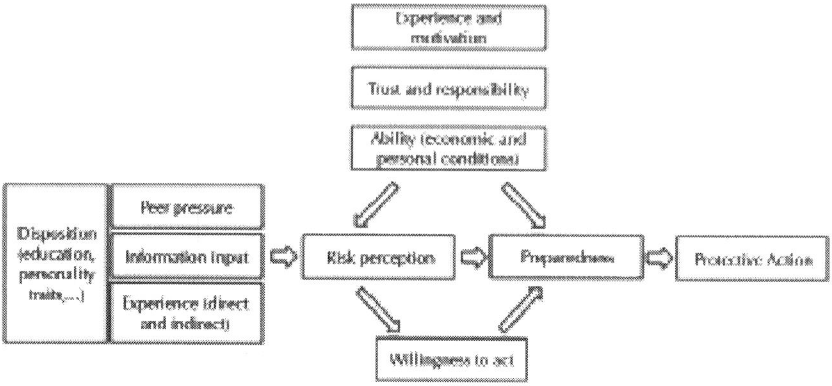

Figure 3: The hazard to action chain; Redrawn from Wachinger et al.([2012]).

Unlike the models discussed before, this model makes few assumptions about the actual decision-making process. However, the authors hypothesize that high trust in authorities could be seen as a form of heuristic processing. Individuals may trust authorities when they are confronted with complex and unknown hazards which require swift decision-making. Further research is required to show whether this model is also applicable to fire emergencies.

Security Motivation System

RP is also studied from an evolutionary perspective. Understanding the biological side of RP can help to develop theories on human behavior and decision-making. Life threatening events, such as fires, are experienced only rarely. Furthermore, indicators of a potential threat are often not easily detectible or may be ambiguous. The question is how organisms adapt to threats that may not occur in every generation. Woody and Szechtman propose a *security motivation system* (SMS) as part of the central nervous system, designed to adapt the organism to extremely rare life threatening events (Szechtman and Woody [2004]). The SMS detects "subtle indicators of potential threat, to probe the environment for further information about these possible dangers, and to motivate engagement in precautionary behaviors, which also serves to terminate security motivation" (Woody and Szechtman [2011]). The authors postulate that the SMS is represented in hardwired neural circuits and its activation motivates protective actions through increased arousal and vigilance, enhanced detection of threatening cues, and the facilitation of future behavioral responses to such cues (Hinds et al. [2010]). Applied to the situation of fires, cues such as the smell of smoke or other people moving to an emergency exit may activate the SMS.

The SMS can be integrated into other theoretical concepts. The SMS has two major functions: (1) to detect and process threat relevant cues and (2) to trigger protective action when a threat is detected (Trower et al. [1990]; Woody and Szechtman [2013]). These functions correspond closely to the assumption about the pre-decisional processes in PADM or the risk-as-feeling approach. The highly automated functions of the SMS can be seen as precursors of the decisional processes in PADM. Woody and Szechtman applied the SMS to policy making (mainly dealing with information about terrorist threats) (Woody and Szechtman [2013]). The authors argue that the SMS is triggered by information about life-threatening events and not by abstract threats, regardless of the actual probability of the event. This offers a potential explanation for cognitive biases or the use of heuristics in emergency situations.

The Mediator Hypothesis

All theoretical frameworks reviewed so far, assume that RP is in some way causally related to evacuation decisions (See also Section What role does perceived risk play in building fire evacuation?). However, evacuation decision-making could be possible without feeling at risk. According to the aforementioned *mediator hypothesis,* RP mediates between individual factors and actual risk reduction behavior. The idea of this hypothesis is that an evacuation decision is not solely dependent on the outcome of an RP process. In fact, it assumes that evacuation is also possible without perceiving risk. Some occupants may evacuate simply because they received instructions to do so. For example, occupants who experienced several fire drills previously may follow instructions to evacuate assuming that these are part of yet another drill and not perceive fire related risk.

Some studies support the assumptions of direct and indirect pathways between individual factors and evacuation decisions. In a study on wildfire risk behavior (note that this is not immediate evacuation behavior), Martin et al. ([2009]) found that risk reducing behavior during wildfires was associated with individual factors such as subjective knowledge and locus of responsibility mediated by perceived risk. In the same study, self-efficacy, defined by Bandura ([1997]) as the extent or strength of one's belief in one's own ability to complete tasks and reach goals, had a direct (non-mediated) effect on risk reduction behavior. In addition, perceived risk was clearly associated with risk reduction behavior (Martin et al. [2009]). Another study found that perceived risk mediated the effect of gender on evacuation from flood but not wind events (Bateman and Edwards [2002]). Future studies should aim to disentangle the causal relationships between perceived risk and evacuation decision-making.

WHAT ROLE DOES PERCEIVED RISK PLAY IN BUILDING FIRE EVACUATION?

In this section literature is presented on how RP affects evacuation behavior and on factors influencing RP itself. Models, such as the PADM discussed in the previous section, state that occupants need to appraise whether a situation provides a threat before they decide to take protective action. However, one may also speculate that RP is not necessarily a precursor of evacuation and that there may be cases in which occupants begin egress without necessarily feeling at risk as the mediator hypothesis suggests. With that said, there are several hypothetically possible links between RP and a protective action decision:

- RP directly causes protective action decision-making and behavior. High perceived risk leads people to engage in protective actions (e.g., evacuation or defend in place), whereas low perceived risk may even lead to non-protective actions (e.g., delaying, actively ignoring cues).

- RP may affect evacuation decision-making and behavior but other factors do so as well.
- RP is a mediator and it accounts for the relationship between a predictor variable (e.g., other human factors) and protective action decision-making (See Section The mediator hypothesis).
- RP is a moderator and it affects the direction and/or strength of the relationship between a predictor variable and protective action.
- RP is a correlate of protective action decision-making and behavior but not a causal factor.
- RP may be independent of evacuation decision-making (i.e., occupants may feel not at risk and evacuate or feel at risk and not evacuate).

Although some of these potential links are mutually exclusive, it is possible that different links are operating in parallel or at different stages of the evacuation process (e.g., in the pre-evacuation and the evacuation period). Future research is necessary to identify which of the potential links are the most important interrelations of RP and protective actions.

RP during the World Trade Center Evacuation on September 11, 2001

A significant amount of research on RP and evacuation has been published on the attacks on the World Trade Center (WTC) on September 11, 2001. Several independent studies found that perceived risk was positively correlated with the likelihood to make evacuation decisions and faster response times, and low perceived risk was associated with delayed evacuation (Day et al.[2013]; Gershon et al. [2007]; Kuligowski [2011]; Sherman et al. [2011]). Kuligowski ([2011]) developed a predictive model of evacuation decision-making based on qualitative interviews with evacuees from the WTC on September 11, 2001. The author found that perceived risk fluctuated throughout the evacuation process (Kuligowski [2011]). Based on the PADM, RP, in the form of risk identification and assessment, was found to play an important role in predicting protective action identification, assessment and implementation (i.e., the decision to evacuate). Some building occupants felt at risk immediately after they noticed initial cues of the impact in WTC 1 (*rapid perceivers*), causing them to decide earlier than others to evacuate. However, in some cases, the perceived risk of the rapid perceivers decreased after a while, as they received additional cues that the author categorized as less intense. Other occupants who received less intense cues, or more intense cues later on in the event, felt at risk only at a later point (*late perceivers*) or not at all (*non-perceivers*). RP was only found to trigger evacuation, if the risk was personalized. Gershon et al. ([2012]) reported that 70% of the interviewed WTC occupants stated that feeling at risk triggered their *evacuation decision*. That is, these results support hypothesis 2 stating that RP may be one important but not the only factor influencing the decision to evacuate.

A closer look reveals an even more complex situation as some of the mentioned studies on the 2001 WTC disaster also found seemingly contradictory results. Sherman et al. ([2011]) studied*evacuation delays* during the attack on the WTC on September 11, 2001. The authors operationalized perceived risk in a single item rating as the perceived "seriousness" of the situation. Here, higher perceived risk was found to be connected to shorter evacuation delays and fewer pre-evacuation behaviors (which again shortened evacuation delays, in line with hypothesis 1, 2, or 5). Sherman et al. ([2011]) also found that higher perceived risk may also lead to more information seeking behavior which in turn prolonged evacuation delays (Hypothesis 3 or 4). In another study on the evacuation from WTC on September 11, 2001 by Kuligowski and Mileti ([2009]), RP was operationalized as a yes/no question on whether occupants believed that somebody else had been killed in the event. Here, higher perceived risk was correlated with more information seeking, more pre-evacuation actions, and longer evacuation delays. However, the path-analysis performed in that study revealed that RP was not directly connected to evacuation delays after controlling for the number of cues, the floor level, information seeking, and the number of pre-evacuation actions (Hypothesis 3). In that model, RP predicted information seeking behavior and the number of pre-evacuation actions in WTC2 (Kuligowski and Mileti [2009]). Sherman et al. ([2011]) suggest a curvilinear relation of RP and information seeking behavior, i.e., if the perceived risk is either extremely high or low, occupants are less likely to seek more information (see also heuristic-systematic approach). Note that differences in the operationalization of RP, data collection and the samples may explain the differences between the studies. The data presented in Sherman et al. ([2011]) was collected several months later than the data from Kuligowski and Mileti ([2009]), and the sub-sample reporting the highest perceived risk in Sherman et al. ([2011]) was not included in Kuligowski and Mileti ([2009]). Kuligowski and Mileti report that perceived risk was rated at four points in a narrative interview process: when the first cues at WTC 1 impact where noticed, at the point the decision to evacuate was made, when participants noticed the WTC 2 impact, and when participants realized that WTC 2 had collapsed. Sherman et al. ([2011]) used a questionnaire format (paper pencil and web based) and the rating of perceived risk was administered independent of a narrative flow. With regard to the definition of RP given earlier, it is not clear to what extent the items used in both studies measured perceived risk. Whereas the item in Sherman et al. ([2011]) could be understood as an evaluation of one's own vulnerability, Kuligowski and Mileti's rating could have been understood as an implicit measure of perceived probability. These differences underline the importance of a clear definition of RP and standardized, reliable, and valid (construct validity, Cronbach and Meehl [1955]) measures of RP.

Another indicator of the complexity of the problem was demonstrated in a study by Day et al. ([2013]) on RP during evacuation from WTC on September 11, 2001. In this study, participants were asked to rate their perceived risk on a seven point Likert scale during different stages of the evacuation process. The authors found a significant negative correlation between perceived risk and response time (Hypothesis 5). However, they also reported that several participants did not give ratings of perceived risks, as they reported not remembering to have evaluated their risk (Hypothesis 6) (Day et al. [2013]). Note that simply not remembering having assessed the risk of a situation does not imply that these occupants did not feel at risk. It is possible, for example, that memory effects biased the participants' responses. Nonetheless, these findings indicate that at least two of the possible links between RP and protective actions mentioned earlier – correlation (5) or independent (6) – are possible.

In summary, the studies discussed here draw a complex picture of the role RP during evacuation from the WTC on September 11, 2001. Some of the differences in the results of the studies may be attributed to the fact that the studies by Sherman et al. ([2011]), Kuligowski and

Mileti ([2009]), as well as Day et al. ([2013]) used different data sets and operationalized RP differently. Given the retrospective correlational nature of the studies, however, an investigation of a potential causal relationship between perceived risk and evacuation behavior is not possible.

Further Evidence and Open Questions

Although the role of RP during evacuation is still inconclusive, one may speculate how RP affects evacuation decision-making. McGee and Russell found that the personalization of risk is an important link between awareness of a hazard and mitigation actions (McGee and Russell [2003]). This finding is in line with the theoretical framework models discussed above (e.g., PADM). According to the heuristic-systematic modeling approach, the level of perceived personal risk affects the level of information processing. Systematic processing can be most likely expected at moderate levels of perceived risk; whereas heuristic processing is expected at either low or high levels. As already mentioned, this may then determine how occupants move through the decisional processes suggested in the PADM.

The question still remains as to how significant RP is to evacuation behavior. Or more specifically, how much variance in evacuation behavior can be explained by RP? For example, Riad et al. ([1999]) found a number of highly significant correlative associations between RP and evacuation (in line with Hypothesis 5), but the reported effect sizes were relatively small. All in all, perceived risk improved the prediction whether someone would evacuate by 8% compared to chance (Riad et al. [1999]). Unfortunately, not many studies on RP and evacuation report effect sizes (See Table 1).

Wachinger et al. ([2012]) propose three hypotheses to explain why some studies on natural disasters did not find a connection between RP and protective action. All three hypotheses introduce moderator variables which also seem potentially applicable to building evacuation. Although the meaning assigned to RP during evacuation from latent threats, such as disasters, may be different, it is possible to develop similar hypotheses for building fire evacuation:

- Occupants perceive high risks but do not decide to engage in protective action because they think that staying in place outweighs the estimated subjective costs of protective action (e.g., having to stop one's work, not wanting to make a fool of one self, or expecting difficulties while evacuating). This could be of particular importance if there are competing motives (e.g., one's own safety vs. property attachment).

- Occupants perceive high risks but trust that others, foe example authorities, will help them. This hypothesis may be less relevant for building evacuation from an acute threat. However, it underlines the importance of credible evacuation communication by authorities.
- Occupants perceive high risks but do not think they have sufficient resources to engage into protective actions (e.g., mobility impaired occupants may not be able to use stairs for evacuation). Again this underlines the importance of credible evacuation communication and/or instructions. Occupants need to know their options in order to engage in protective action.

Future studies are clearly necessary to understand the role of RP in building fire evacuation. These studies should investigate (1) the causal relationship between RP and evacuation decision-making and behavior, as well as (2) underlying reasons why some occupants do not evacuate although they may feel at risk.

Factors Potentially Modulating RP

Several factors potentially modulate RP which can be broadly differentiated into situational, individual, social, and organizational factors (Table 2). Some of these factors are dynamic, in that they may change during the course of an emergency situation, (e.g., available fire cues or emotional states) and some are static (e.g., context or previous experience). Furthermore, these factors interact with each other and can be affected by the RP process itself. For some of the static factors, distributions can be either assumed or derived from the literature. For example, the factor*prior experience* could be operationalized as the percentage of occupants in a building who have previously experienced an event or evacuation.

Table 2: Current knowledge on factors affecting perceived risk and evacuation behavior

Factor	Category	Static/ dynamic1	Effect on perceived risk
Fire Cues	Situational	Dynamic	More, closer, unexpected and more intense fire cues lead to higher perceived risk
Hazard proximity	Situational	Dynamic	Inconclusive
Floor level	Situational	Dynamic	The higher the floor, the higher the perceived risk
Context	Situational	Static	Inconclusive
Credibility of information	Situational	Static	Credibility of information moderates information processing and perceived risk with potential interaction effects of the source of information (another person vs. system)
Complexity of information	Situational	Dynamic	Inconclusive
Gender	Individual	Static	Tendency toward lower perceived risk in men, but effects are potentially modulated by age and context
Age	Individual	Static	Inconclusive
Previous experience	Individual	Static	Direct effects of previous experience on perceived risk are inconclusive.

Behavioral training	Individual	Static	Inconclusive
Hazard knowledge	Individual	Static	Knowledge about hazards increases perceived risk
Property attachment	Individual	Static	Inconclusive
Personality traits	Individual	Static	Inconclusive
Emotional states	Individual	Dynamic	High arousal and state anxiety increase perceived risk
Medical factors	Individual	Dynamic	Inconclusive
Cognitive abilities	Individual	Static	Inconclusive
Information Processing	Individual	Dynamic	Information that is processed easily may be associated with lower perceived risk
Trust in authorities	Individual	Static	High trust reduces perceived risk; low trust increases perceived risk
Cognitive bias	Individual	Dynamic	Inconclusive
Behavior of others	Social	Dynamic	Behavior of others moderates the link between perceived risk and protective action
Social roles	Social	Dynamic	Inconclusive
Groups	Social	Dynamic	Higher perceived risk in groups
Organizational context	Organizational	Dynamic	Inconclusive

Note:[1]Dynamic factors can change in the course of a fire emergency, e.g., the number of fire cues may increase or decrease with time.

References for the findings are given in the text in Section 4.4.1 to 4.4.4.

As the role of RP during fire evacuation is complex (see above), the factors potentially influencing RP also interact with each other and affect other important variables in the evacuation process. For example, the number and intensity of cues and the floor level affected not only RP, but also had a direct impact on pre-evacuation delays in two studies on the WTC evacuation on September 11, 2001 (Kuligowski and Mileti [2009]; Sherman et al. [2011]). The exact interaction among RP, evacuation decision-making and evacuation delay is still not entirely clear (e.g., what is a mediating or a moderating variable? Which factors are mainly correlates but have effects on evacuation behavior independent of RP?). Findings on several of the factors identified as being connected to RP and evacuation are described below.

Situational Factors

Situational factors refer to all aspects of circumstances at a given moment that influence RP and/or evacuation. These cues originate mainly from the physical environment of an occupant.

Fire cues refer to all cues initiated by a fire[1]. Fire cues that are greater (in number), closer in proximity, and more intense have been linked to higher perceived risk (Sherman et al. [2011]; Kuligowski and Mileti [2009]; Kuligowski [2011]). In addition, sudden and unexpected cues may increase perceived risk in the sense that unusual, surprising events produce cues which occupants cannot identify immediately (Sherman et al. [2011]).The accuracy of information conveyed by fire cues is relevant, especially since information that clearly and unambiguously indicates threat can increase perceived risk. This may explain why studies found that poorly designed alarm systems may not induce high enough perceived risk in occupants (Caroly et al. [2013]; Kuligowski [2011]).

Hazard proximity refers to the spatial proximity of occupants to a threat; its role on RP is still inconclusive although research suggests that it may modulate perceived risk. Some studies found that the higher the perceived risk, the closer the hazard was to occupants (Kuligowski and Mileti[2009]), while other studies did not find this effect (Fahy and Proulx [2002]). It is possible that other factors, such as the relative location of occupants to the hazard and known exit routes modulate the effect of hazard proximity. Visibility, vertical and horizontal distance, or other factors, may be important confounding factors. In addition, it is not clear if the relation between hazard proximity and perceived risk is, for example, linear or non-linear.

Floor level refers to the absolute floor level of an occupant in a building, irrespective of his/her distance to the fire (which would be measured by the *hazard proximity* factor mentioned above). The current state of research concludes that perceived risk increases with floor level in high-rise buildings. In the case of a full building evacuation, the *floor level* in a high-rise building was a significant predictor of perceived risk during the evacuation from the WTC on September 11, 2001 (the higher the floor, the more perceived risk) in one study but not in another (Sherman et al.[2011]; Kuligowski and Mileti [2009]). Although future studies should investigate whether the absolute floor level or the position relative to the fire origin (*hazard proximity*) is more relevant.

Context, broadly defined as the general circumstances of an event, affects human behavior in fire in several ways and its effect on RP is still not fully understood. Preparedness, vigilance, and the interpretation of fire cues may vary over different contexts (e.g., public events, workplace, or home setting). A questionnaire study demonstrated that participants' self-reported perceived risk varied over different settings (e.g., financial vs. safety related decision-making) (Weber et al.[2002]). The underlying mechanism may be that environmental cues are interpreted differently over different contexts. In a residential home, for example, occupants may perceive cues about a fire from the fire itself or from smoke detectors. In public buildings most occupants may only receive information from the fire alarm system (e.g. through public announcements). Another explanation may be that cognitive biases (e.g., normalcy bias), social roles and perceived responsibility, availability of emergency procedures (e.g. evacuation plans), and the interpretation of cues may vary across contexts.

Credibility of information[b] refers to the perceived level of credibility that a person assigns to a piece of information or source of information. Overall, this factor moderates information processing and RP with potential interaction effects of the source of information. Credibility of risk-related messages affects information processing (see, heuristic systematic approach) as well as RP and has been extensively studied in the context of disaster preparedness (Mileti and Sorensen[1990]). One study on long term RP showed that risk assessment after receiving risk-related information was in part mediated by heuristic and systematic information processing. Here, highly credible sources were associated with more heuristic information processing and lower perceived risk. In turn, low credibility of an information source is associated with more systematic information processing and higher perceived risk. If the source of information was an industry or government organization, higher credibility was correlated with lower perceived risk scores in this study. If the source of information was another person, however, this correlation was inverted (Trumbo and McComas [2003]). Another study on hurricane evacuation showed that people use different sources of information and their trust in the credibility of these sources varies (Lindell and Whitney [2000]).

[b]Here, the authors are describing the effect of information credibility on risk perception. Whereas this is inconclusive, other research, as well as NIST guidance, suggests that credibility of information has an important influence on evacuation decision-making, response, and protective action behavior (Kuligowski and Omori [2014]).

The *complexity* of a situation, including the information provided to building occupants in a fire situation, may affect whether information is processed systematically or heuristically. In one basic research study, participants rated ostensible food additives as more harmful when their names were more difficult to pronounce than when their names were easier to pronounce. The study indicates that information which is more demanding to process increases perceived risk (Song and Schwarz [2009]). Transferring that to evacuation scenarios, Drabek hypothesizes that inconsistency, ambiguity and overload of information increase emergent perceived risk (Drabek[2001]). In line with these findings, another study found that mobility impaired building occupants associated lack of information (ambiguity) with highest ratings of concern during a fire evacuation (McConnell and Boyce [2013]). However, given the small number of studies on the effect of the complexity of a situation on RP, further research is clearly necessary.

Individual Factors

Individual factors refer to factors within a person that may affect RP and evacuation behavior. These can be either state (i.e., dynamic, for example, emotional states or arousal) or trait (i.e., stable, for example, gender, age, cognitive abilities) variables.

Gender

Lower perceived risk (Slovic [2010a]) and less risk-averse attitudes (Weber et al. [2002]) of men compared to women might explain gender differences in evacuation behavior. However, no influence of gender on risk identification and assessment was found in Kuligowski's analysis of evacuation decision-making during the WTC disaster on September 11, 2001 (p. 148) (Kuligowski[2011]). In Sherman et al.'s WTC evacuation study, being female was associated with increased perceived risk (Sherman et al. [2011]). A meta-analysis found that men were more likely to engage in risk taking behavior, but that this effect was modulated by context (i.e., the kind of threat) and age (i.e., with growing age, the differences seemed to get smaller) (Byrnes et al.[1999]). In another study, gender differences in RP could be explained by differences in self-reported fear and anger (Lerner et al. [2003]). In summary, men seem to perceive less risk than women.

Age is correlated with several evacuation-relevant variables (e.g., experience, cognitive and physical abilities, education, social role, etc.); however, its role with regard to RP is still inconclusive. Some authors argue that older adults are better in risk evaluation than younger adults since they have to practice risk-related decisions more frequently in their daily life (e.g., medication labeling, adaption to changes in physical fitness) (Wilson et al. [2013]; McLaughlin and Mayhorn [2014]). This is in line with research on driving behavior, which states that deficits due to reduced physical abilities or reaction times can be compensated on a strategic and tactical level (higher vigilance). Further research is needed since some studies found that older occupants are less likely to evacuate (Riad et al. [1999]) but others found no relation between age, perceived risk, and evacuation delay (Sherman et al. [2011]; Kuligowski and Mileti [2009]; Kuligowski[2011]).

Previous experience with fire emergencies or similar situations may significantly affect RP, vigilance, and preparedness and has been identified as one of the strongest predictors of increased perceived risk during natural disasters (Wachinger et al. [2012]). However, experiencing a disaster without experiencing personal harm may decrease perceived risk. (For an overview of studies on RP, experience, and natural disasters, see Wachinger et al. [2012]). Research from volcano disasters showed that having experience in a disaster diminished differences in RP between volcano experts and untrained participants (Bird and Gisladottir [2012]). Similarly, survivors of the 1993 WTC bombing had shorter evacuation delays than occupants who had no such experience during the evacuation of the WTC on September 11, 2001 (Day et al. [2013]). Therefore, the effects of previous experience on RP are still under debate, although it seems possible that increased perceived risk moderates the connection between evacuation decision-making and previous experience.

Behavioral training aims to convey behavioral or theoretical knowledge through practice. Although the effects of behavioral training on RP is still unclear, it is known to improve evacuation behavior (e.g., Kinateder et al. [2013]). The ability of novice drivers to detect hazards can be improved through training, as one study showed that trained participants scanned the environment for potential threats more frequently and efficiently than the control group (Pradhan et al. [2006]). That is, training may increase *preparedness* and *vigilance* for fire cues, and the effectiveness of training depends on the severity of perceived risks (Burke et al. [2011a]).

Hazard knowledge refers to the knowledge that any person has related to specific types of hazards associated with an incident, including the consequences of the hazard and appropriate responses. This factor has been shown to increase perceived risk, although these effects are complex and still not fully understood. In line with studies showing that knowledge is correlated with the adoption of risk reduction behaviors (Lindell and Whitney [2000]; Shields et al. [2009]), Kuligowski and Mileti found that obtaining additional information after receiving initial fire cues was weakly correlated, but with statistical significance, with perceived risk during the evacuation from WTC on September 11, 2001 (Kuligowski and Mileti [2009]). However, unknown or ambiguous events are also associated with increased perceived risk (Song and Schwarz [2009]). In turn, *familiarity* with an event reduces perceived risk (Riley [2014]), although familiarity does not necessarily imply knowledge. An overuse of warnings and false alarms may consequently lead to a desensitization of occupants and may reduce their perceived risk during a real emergency (Rando et al. [2007]). Further research is necessary to disentangle the effects of hazard knowledge and familiarity on RP.

Property attachment or *territorial functioning* may not directly affect RP, but may mitigate the connection between perceived risk and evacuation (see also *Context*). In studies on hurricane evacuation, homeowners reported that a reason for not evacuating was the fear of looting (i.e., perception of risk to personal property; Riad et al. [1999]; Kang et al. [2007]; Huang et al.[2012]). In some cases, occupants returned to their desk to pick up personal items during evacuation of WTC on September 11, 2001 (Kuligowski [2011]). Further research is necessary to clarify the effects of property attachment on RP.

Personality traits refer to relatively stable "patterns of thoughts, feelings, and actions in a diverse array of psychological phenomena, including motives, wishes, apperceptions [c], and attitudes, as well as behaviors in which a person processes information (Mccrae and Costa [1995])." Although future studies need to clarify the exact role of personality traits and RP during fire emergencies, some personality traits, such as impulsivity or sensation seeking, are related to risk taking behavior and may be important for RP (Zuckerman and Kuhlman [2000]; Ryb et al. [2006]). Based on personality traits, individuals may vary in how they perceive risk. Highly impulsive occupants, for example, may require a lower number of fire cues to perceive a high enough risk before they decide to take protective actions. One study found that the relation between personality traits and risky driving behavior was mediated by risk-related attitudes (Ulleberg and Rundmo [2003]).

[c]Apperception in the psychology literature refers to consistent patterns on how people perceive their environment in relation to their past experience.

Emotional states, such as state anxiety, are correlated with arousal (the activation of the sympathetic nervous system), and can increase perceived risk. Higher arousal is associated with more impulsive information-processing (Strack and Deutsch [2004]) and may bias RP. High state anxiety affects the way that hazard cues are processed and reduces cognitive resources (Mathews and MacLeod [1985]; Yiend [2010]). Emotional states may also affect the readiness with which cues are interpreted as threatening and an attentional bias on threatening stimuli (Cisler and Koster [2010]). One questionnaire study on terrorism-related hazards found that risk appraisals were modulated by fear and anger. Highly fearful participants reported higher perceived risk, and participants scoring high on the anger scale also reported lower perceived risk (Lerner et al.[2003]).

Medical factors (including intoxication) affect how fire cues are perceived and information is processed, but the effects on RP are still inconclusive. For example, alcohol intake distorts RP in the sense that that it modulates arousal and may lead to more risky behavior (Mongrain and Standing [1989]). However, the range of medical factors potentially modulating perceived risk is vast and future studies are necessary.

Cognitive abilities refer to the ability to understand fire related cues. Although still inconclusive, some cognitive impairments, such as mild cognitive impairment or dementia (Christensen et al.[2013]; Brown et al. [2012]) may reduce the ability to perceive and understand fire related cues and reduce the ability to perceive risk as well as to comply with evacuation procedures. So far, there are no studies that directly address RP and cognitive abilities in the context of evacuation and future studies are necessary to understand their role during evacuation.

Information processing and RP potentially interact; however, the exact relation is still inconclusive. Basic research shows that low processing fluency (i.e., the ease with which information can be processed) fosters the impression that a stimulus is unfamiliar, which in turn results in perceptions of higher risk (Song and Schwarz [2009]). That is, higher cognitive load when processing unfamiliar information is associated with higher perceived risk. Empirical studies on information processing and RP during fire evacuation are necessary to verify whether these results can be transferred to emergency situations.

High *trust in authorities* reduces perceived risk (and may even lead occupants to underestimate a hazard in unprotected areas), whereas low trust (or distrust) increases perceived risk (Wachinger et al. [2012]). The authors further note that trust in authorities can be understood as a heuristic supporting decision-making in complex situations and when facing unknown or ambiguous threats. Trust in authorities, similar to credibility of information (see above), may also mediate the path between perceived risk and protective actions. Furthermore the authors hypothesize that damaging trust may increase perceived risk (Wachinger et al. [2012]). Another hypothetical scenario study found a correlation between the degree of trust in authorities and hazard appraisals for certain technological risks (Siegrist and Cvetkovich

[2000]). As most of the research on trust in authorities and RP focusses on evacuation from natural disasters, further research with regard to fire evacuation is necessary.

Cognitive biases refer to systematic distortions in human information processing and decision-making. The use of heuristics may lead to such biases.

General RP bias: In general, perceived risk of events is correlated with the actual risk. However, there are some biases in the sense that small risks are overestimated and high risks are underestimated (Sjöberg [2000]; Lichtenstein et al. [1978]; Thompson and Mingay [1991]; Mbaye and Kouabenan [2013]).

Positivity bias (comparative optimism) refers to the fact that occupants consistently rate their own personal risk as lower than the risk to others. This phenomenon is well documented in the literature, and was found in one study on tunnel fire emergencies (Sjöberg [2000]; Horney et al.[2010]; Mbaye and Kouabenan [2013]).

Locus of Control/Perceived Control: risks perceived to be under one's own control are more acceptable than risks perceived to be controlled by others. Illusion of control was found to be correlated with perceived invulnerability (positivity bias) and negatively with perceived risk in a study on accidents in chemical and nuclear facilities (Mbaye and Kouabenan [2013]).

Normalcy bias reduces perceived risk and refers to a tendency to attribute cues to 'normal' events during disasters and not to catastrophic events. During the evacuation of the WTC on September 11, 2001 occupants, especially from the lower floors, reported relatively low perceived risk which may be attributed to the assumption that nothing extraordinary was going on in the building (McConnell et al. [2010]).

Social Factors

Social factors mainly refer to the effect of others on one's own RP and behavior. This can be broadly labeled as social influence. Social influence is defined as changes in attitudes, beliefs, opinions or behavior as a result of the fact that one is confronted with the attitudes, beliefs, opinions, or behavior of others (Hewstone and Martin [2008]).

Behavior of Others

The behavior of others potentially moderates the link between RP and protective action. Seeing other occupants evacuate provides a cue for an emergency and may increase personal perceived risk. In turn, passive behavior of others may trigger the normalcy bias (i.e., that nothing is wrong) and reduce perceived risk. Alternatively, social influence may lead occupants to ignore their own appraisal of the situation. Studies from the evacuation of a cinema theater showed that the non-evacuation behavior of others could thwart evacuation (Nilsson and Johansson [2009]; Kinateder et al. [2014a], [b]). Social influence on RP may be a function of knowledge, as one study showed that experts, in comparison to untrained participants, rely less on information derived from others (Siegrist and Cvetkovich [2000]). Further studies testing the specific relationship of social influence, perceived risk and protective action are necessary.

The effect of *social roles* on RP is still inconclusive. However, it is possible to assume that social roles affect RP as a function of perceived responsibility and knowledge. For example, trained fire wardens may improve their hazard detection skills and be more vigilant. In turn, the behaviors of fire wardens, or occupants with assigned authority, may influence the perceived risk of other occupants. Survivors of the WTC attacks reported that being told to evacuate by others (especially people in fire safety roles or roles of authority) triggered their evacuation decision (See also *Social trust*) (Gershon et al. [2012]; Kuligowski [2011]). Further studies are also necessary to investigate the impact social roles have on the scope of an occupants' perceived risk (See 3.1 Scope of RP). For example, one may speculate that occupants whose social role includes high perceived responsibility for others (e.g., a fire fighter or a parent) extend the scope of their RP to others.

Occupants in *groups* may experience higher perceived risk than occupants who are alone during a fire emergency. During the evacuation from WTC 1 on September 11, 2001, occupants who grouped together during the event reported higher perceived risk (Trumbo and McComas [2003]). However, only one study was found to provide evidence of this linkage. Therefore, future studies are necessary to test if occupants with higher perceived risk are more likely to form groups, or whether forming groups increases perceived risk.

Organizational Factors

Organizational factors refer to the effects of the organizational structure on RP during evacuation. In a study on the September 11, 2001 WTC evacuation, participants working for the New York/New Jersey Port Authority reported higher perceived risk during the incident (Sherman et al. [2011]). One may speculate that an organization's safety climate and culture affects RP and social roles, which in turn influences protective action of the organization's member. A qualitative study of the WTC disaster on September 11, 2001 showed that evacuation was affected by worksite preparedness planning, including the training and education of building occupants as well as risk communication (Gershon et al. [2007]).

Organizational Context

Depending on where an emergency occurs (e.g., work, home, public places; see also *context*), RP and evacuation behavior may be different. In one questionnaire study participants reported higher compliance rates to hypothetical evacuation orders if they were at work than at home (Dombroski et al. [2006]). One reason might be that occupants felt safer or perceived less risk in their home environment. Depending on the organizational context, the perceived risk necessary before an evacuation decision is made might be different.

Summary of Factors

Table 2 summarizes the findings regarding the individual factors potentially affecting RP during emergencies. A literature review on RP during natural hazards (e.g., evacuation from hurricanes or floods) concluded that the previous experience and lack of trust in authorities had the strongest direct effects on RP (Wachinger et al. [2012]). Future studies are necessary to test whether this holds true for building fire evacuations as well. Similarly, further research is clearly necessary regarding all the factors identified in the present review, since so many of the relationships were inconclusive (See Table 2).

OVERVIEW OF STUDIES

Table 1 gives an overview of studies used in this review. The studies are sorted according to their relevance to RP during building fire evacuation. A comparison of the studies reveals that there are very few studies on RP in the context of a building fire evacuation.

LIMITATIONS

There are some limitations to the present review. The literature reviewed for this report varies significantly in nature and scope. The question of scope, or what to include and what not to include in this review, was therefore an important one (Ogilvie et al. [2005a], [b]). The theoretical models selected and discussed here focus on RP as the process of an individual occupant. There are additional theories that address risk and RP in other contexts, e.g. in strategic decision-making (e.g., Zheng and Cheng [2011]; Lo et al. [2006]) or as a social or organizational phenomenon such as the social amplification of risk framework (Kasperson et al. [1988]). However, these theories are beyond the scope of the present article.

Another issue is publication bias. Generally speaking, very few studies on RP report results where no correlations were found between RP and evacuation. This indicates that there might be a publication bias towards positive relations between RP and evacuation. The Cochrane collaboration and other researchers have repeatedly shown that studies with significant and positive results are easier to find than those with non-significant or 'negative' or null results (Guyatt et al. [2011]). This may have caused an over-representation of studies finding correlations between RP and evacuation.

As already mentioned previously, the bulk of the literature on evacuation and RP relies on self-reported rating scales. Many authors operationalized RP with 1 item questions such as "How 'at risk' did you feel at particular moments during the evacuation process?" (e.g., Kuligowski [2011]; Day et al. [2013]; Martin et al. [2009]; Siebeneck and Cova [2012]; Matyas et al. [2011]; McConnell et al. [2010]; Horney et al. [2010]). Single items are an easy to use and economical approach to measure RP. However, the question(s) may not grasp RP in its full complexity and it is possible that participants had different concepts about what they meant when they rated their perceived risk. Using different methods of measurements may lead to significantly different outcomes, as demonstrated by the comparison of two studies on evacuation from the WTC on September 11, 2001 (Sherman et al. [2011]; Kuligowski and Mileti [2009]). Furthermore, it is crucial that the measurement tool actually reflect the construct of interest (construct validity). It is questionable if a single item can validly measure a construct that consists of two independent dimensions (perceived probability and vulnerability). However, single item measures can have sufficient predictive validity (Wanous et al. [1997]; Robins et al. [2001]; Elo et al. [2003]). It is necessary to develop and test measuring tools (e.g., questionnaires) for RP during building fire evacuations that meet common quality criteria for objectivity, reliability, and validity (Eignor[2013]) of psychometric testing.

Most of the data about RP and evacuation reported here relies on self-report data. As previously noted, the understanding of the term risk perception may vary greatly within the population. Self-report studies are extremely useful to get an understanding of occupants' experiences and behaviors during evacuation. However, self-report data are prone to bias due to social desirability and other sources of bias (e.g., memory effects). That is, self-reported behavior or behavioral intentions may differ greatly from actual behavior. Slovic ([2010a]) reports the case of a study in which participants were asked whether the construction of a nuclear power plant would stop them from using an adjacent beach. Most participants stated that they would stop using the beach if the plant was built. The power plant was built, and no decline in the attendance of the beach was observed.

In the present review, the authors discussed the effect of several factors on RP. In each case, an attempt was made to identify correlations or causal effects between each factor and RP, individually. However, RP in complex situations, like a fire emergency, is most likely determined by multiple variables, which may interact with each other. That is, the conclusions of the present review may oversimplify how various factors increase or decrease perceived risk and may neglect potential interaction effects between the factors.

The present review on the role of RP during fire evacuation is heavily based on studies of one single event (i.e., the attacks on WTC on September 11, 2001). Although these studies revealed comparable results using independent databases, and knowledge on RP has significantly advanced based upon these research efforts, it stands to reason that the events of September 11, 2001 may not allow for generalization to all other building fires. Future studies are necessary to build a broader database. Such studies should take into account different contexts (e.g., with regard to occupancy or location). Only a limited number of studies were found using data from laboratory settings or drills. Additionally, prospective studies are extremely scarce in this field (Table 1). The development of ecologically valid and ethical laboratory paradigms for the study of evacuation and RP may prove especially useful.

Finally, this literature review depended on the accessibility of sources. This review is limited to the libraries of the National Institute of Standards and Technology and the University of Würzburg, Germany. For literature without full text access from either of these two libraries or through interlibrary loan, abstracts were considered, or the source was ignored.

CONCLUSIONS AND OUTLOOK

The first goal of this overview was to clarify the concept of RP in the context of building fire evacuation and to provide a definition of RP specifically for this field. RP was defined and differentiated from other similar concepts, such as situation awareness. In this paper, RP is seen as a psychological process comprising the subjective evaluation of the probability to be affected by an imminent threat and an assessment of one's own perceived vulnerability and coping resources. It is modulated by affects and prone to cognitive biases. In a second step, the following relevant theoretical frameworks on RP from evacuation research were identified and described: Heuristic-systematic approach, PADM, Transactional stress model, reasoned action models, and SMS. We believe that this synopsis may contribute to theory development in the field of evacuation research.

In a next step, factors potentially influencing RP during building fire evacuation were identified and discussed. The results of this discussion, summarized in Table 2, revealed that the number of fire cues, floor level in high-rise buildings, credibility of information, gender, previous experience, hazard knowledge, certain emotional states, information processing, certain cognitive biases, the behavior of others, and groups can affect RP.

Future research will have to clarify the relationship of the factors identified in the present review. Specifically, three future research steps are necessary: (1) Development of a self-report questionnaire of RP for fire evacuation that meets common quality criteria of psychometric testing. The variety in which perceived risk was measured in the studies reviewed in the present article indicates that a common standard to study RP during fire evacuation is necessary. An objective, reliable, and valid questionnaire is necessary to understand RP during fire evacuation. (2) Identification of specific effects of perceived risk during the pre-alarm and the protective action phase of a fire emergency. Controlled laboratory studies which systematically manipulate RP could shed light on how perceived risk influences RSET. One possibility could be to manipulate the expectation and emotion component of perceived risk through conditioning (e.g., by presenting aversive stimuli combined with a scenario (context information)) or priming (e.g., by presenting information about fire emergencies) and then confront participants with a hypothetical emergency scenario. Another possibility would be to compare the reaction to fire cues of participants who differ systematically in trait RP (e.g., by comparing reactions of highly fearful and non-fearful participants). (3) Development of a holistic predictive model on the interaction of the factors potentially modulating RP. Although the present review identified a set of factors that most likely influence perceived risk during fire evacuation, it is unclear how strong the effects of individual factors are and how these factors interact with each other.

The present review demonstrates that RP is relevant to evacuation outcome variables such as evacuation decision-making and evacuation delays. We introduce a definition of RP during fire evacuation, allowing a more precise operationalization of the concept. A precise operationalization of RP potentially allows researchers to explain additional variance in occupants' evacuation decision-making and behavior, and, consequently, may improve the prediction of ASET/RSET in engineering tools.

AUTHORS' CONTRIBUTION

MK designed the study, performed the literature search, discussed the data, performed the review and drafted the manuscript. EK participated in the design of the study, discussed the data and co-drafted and revised the manuscript. PR participated in the design and coordination of the study and helped to draft the manuscript. All authors read and approved the final manuscript. RP discussed the data and revised the manuscript for important intellectual content conceived of the study, and participated in its design and coordination and helped to draft the manuscript. All authors read and approved the final manuscript.

ACKNOWLEDGEMENTS

The authors would like to thank Steve Gwynne, Anthony Hamins, Kathryn Butler and Jiann Yang for critically reviewing the manuscript and providing valuable insights.

The study was funded by the National Institute of Standards and Technology.

Parts of the present manuscript will be published in the NIST Technical Note series.

REFERENCES

1. Ajzen I (2011) Theory of planned behavior. Handb Theor Soc Psychol Vol One 1:438
2. Averill JD, Groner N, Kuligowski ED, Mileti D, Peacock R (2007) Federal Building and Fire Safety Investigation of the World Trade Center Disaster: Occupant Behavior, Egress, and Emergency Communications. NIST NCSTAR 1–7
3. Averill JD, Mileti D, Peacock R, Kuligowski E, Groner N, Proulx G, Reneke P, Nelson H (2012) Federal investigation of the evacuation of the World Trade Center on September 11, 2001. Fire Mater 36(5–6):472-480 doi:10.1002/Fam.2162
4. Bandura A (1997) Self-efficacy: The exercise of control. Macmillan

5. Bateman JM, Edwards B (2002) Gender and evacuation: A closer look at why women are more likely to evacuate for hurricanes. Nat Hazards Rev 3(3):107-117
6. Bird DK, Gisladottir G (2012) Residents' attitudes and behaviour before and after the 2010 Eyjafjallajokull eruptions-a case study from southern Iceland. Bull Volcanol 74(6):1263-1279 doi:10.1007/s00445-012-0595-z
7. Brenkert-Smith H, Dickinson KL, Champ PA, Flores N (2013) Social amplification of wildfire risk: the role of social interactions and information sources. Risk Anal : Off Publ Soc Risk Anal 33(5):800-817 doi:10.1111/j.1539-6924.2012.01917.x
8. Brown LM, Dosa DM, Thomas K, Hyer K, Feng Z, Mor V (2012) The effects of evacuation on nursing home residents with dementia. Am J Alzheimers Dis Other Dement 27(6):406-412 doi:10.1177/1533317512454709
9. Burke MJ, Salvador RO, Smith-Crowe K, Chan-Serafin S, Smith A, Sonesh S (2011) The dread factor: how hazards and safety training influence learning and performance. J Appl Psychol 96(1):46-70 doi:10.1037/a0021838
10. Business Dictionary (2013) Definition: risk assessment
11. Byrnes JP, Miller DC, Schafer WD (1999) Gender differences in risk taking: A meta-analysis. Psychol Bull 125(3):367-383 doi:10.1037/0033-2909.125.3.367
12. Caroly S, Kouabenan DR, Gandit M (2013) Analysis of danger management by highway users confronted with a tunnel fire. Saf Sci 60:35-46 doi:10.1016/j.ssci.2013.06.006
13. Cauberghe V, De Pelsmacker P, Janssens W, Dens N (2009) Fear, threat and efficacy in threat appeals: Message involvement as a key mediator to message acceptance. Accid Anal Prev 41(2):276-285 doi:10.1016/j.aap.2008.11.006
14. Chaiken S (1980) Heuristic Versus Systematic Information-Processing and the Use of Source Versus Message Cues in Persuasion. J Pers Soc Psychol 39(5):752-766 doi:10.1037//0022-3514.39.5.752
15. Chaiken S, Eagly AH (1989) Heuristic and Systematic Information Processing within and. Unintended thought 212

16. Chaiken S, Maheswaran D (1994) Heuristic Processing Can Bias Systematic Processing - Effects of Source Credibility, Argument Ambiguity, and Task Importance on Attitude Judgment. J Pers Soc Psychol 66(3):460-473 doi:10.1037/0022-3514.66.3.460
17. Christensen JJ, Richey ED, Castaneda H (2013) Seeking safety: predictors of hurricane evacuation of community-dwelling families affected by Alzheimer's disease or a related disorder in South Florida. Am J Alzheimers Dis Other Dement 28(7):682-692 doi:10.1177/1533317513500837
18. Cisler JM, Koster EH (2010) Mechanisms of attentional biases towards threat in anxiety disorders: An integrative review. Clin Psychol Rev 30(2):203-216 doi:10.1016/j.cpr.2009.11.003
19. Cronbach LJ, Meehl PE (1955) Construct validity in psychological tests. Psychol Bull 52(4):281-302
20. Day RC, Hulse LM, Galea ER (2013) Response Phase Behaviours and Response Time Predictors of the 9/11 World Trade Center Evacuation. Fire Technol 49(3):657-678 doi:10.1007/s10694-012-0282-9
21. Dombroski M, Fischhoff B, Fischbeck P (2006) Predicting emergency evacuation and sheltering behavior: a structured analytical approach. Risk Anal : Off Publ Soc Risk Anal 26(6):1675-1688 doi:10.1111/j.1539-6924.2006.00833.x
22. Drabek TE (2001) Disaster warning and evacuation responses by private business employees. Disasters 25(1):76-94
23. Eignor DR (2013) The standards for educational and psychological testing
24. Elo AL, Leppanen A, Jahkola A (2003) Validity of a single-item measure of stress symptoms. Scand J Work Environ Health 29(6):444-451
25. Endsley MR (1988) Design and evaluation for situation awareness enhancement. In: Proceedings of the Human Factors and Ergonomics Society Annual Meeting, vol 2. SAGE Publications, Thousand Oaks, CA, pp 97–101
26. Endsley MR (1995) Measurement of Situation Awareness in Dynamic-Systems. Hum Factors 37(1):65-84 doi:10.1518/001872095779049499

27. Endsley MR, Jones W (2013) Situation awareness. The Oxford Handbook of Cognitive Engineering:88
28. Fahy RF, Proulx G (2001) Toward creating a database on delay times to start evacuation and walking speeds for use in evacuation modeling
29. Fahy RF, Proulx G (2002) A comparison of the 1993 and 2001 evacuations of the World Trade Center. In: Proceedings of the 2002 Fire Risk and Hazard Assessment Symposium. pp 111–117
30. Finucane ML, Alhakami A, Slovic P, Johnson SM (2000) The affect heuristic in judgments of risks and benefits. J Behav Decis Mak 13(1):1-17 doi:10.1002/(Sici)1099-0771(200001/03)13:1
31. Firing K, Karlsdottir R, Laberg JC (2009) Social influence in military leadership training. Leadersh Org Dev J 30(8):709-721
32. Fischhoff B (1995) Risk Perception and Communication Unplugged: Twenty Years of Process1. Risk Anal 15(2):137-145 doi:10.1111/j.1539-6924.1995.tb00308.x
33. Fishbein M (1979) A theory of reasoned action: some applications and implications
34. Floyd DL, Prentice-Dunn S, Rogers RW (2000) A meta-analysis of research on protection motivation theory. J Appl Soc Psychol 30(2):407-429 doi:10.1111/j.1559-1816.2000.tb02323.x
35. Furnham A, Boo HC (2011) A literature review of the anchoring effect. J Socio-Econ 40(1):35-42 doi:10.1016/j.socec.2010.10.008
36. Gershon RR, Qureshi KA, Rubin MS, Raveis VH (2007) Factors associated with high-rise evacuation: qualitative results from the World Trade Center Evacuation Study. Prehosp Disaster Med 22(3):165-173
37. Gershon RRM, Magda LA, Riley HEM, Sherman MF (2012) The World Trade Center evacuation study: Factors associated with initiation and length of time for evacuation. Fire Mater 36(5–6):481-500 doi:10.1002/Fam.1080
38. Green DM, Swets JA (1966) Signal detection theory and psychophysics. Wiley, New York.
39. Greenberg J, Solomon S, Veeder M, Lyon D, Pyszczynski T, Rosenblatt A, Kirkland S (1990) Evidence for Terror Management Theory.2. The Effects of Mortality Salience on Reactions to Those Who Threaten or Bolster the Cultural Worldview. J Pers Soc Psychol 58(2):308-318 doi:10.1037//0022-3514.58.2.308

40. Greening L, Dollinger SJ, Pitz G (1996) Adolescents' perceived risk and personal experience with natural disasters: an evaluation of cognitive heuristics. Acta Psychol (Amst) 91(1):27-38
41. Griffin MA, Neal A (2000) Perceptions of safety at work: a framework for linking safety climate to safety performance, knowledge, and motivation. J Occup Health Psychol 5(3):347
42. Guyatt GH, Oxman AD, Montori V, Vist G, Kunz R, Brozek J, Alonso-Coello P, Djulbegovic B, Atkins D, Falck-Ytter Y (2011) GRADE guidelines: 5. Rating the quality of evidence—publication bias. J Clin Epidemiol 64(12):1277-1282
43. Heilbrun K, Wolbransky M, Shah S, Kelly R (2010) Risk communication of terrorist acts, natural disasters, and criminal violence: comparing the processes of understanding and responding. Behav Sci Law 28(6):717-729 doi:10.1002/bsl.940
44. Hewstone M, Martin R (2008) Social Influence. In: Hewstone M, Ströbe W, Jonas K (eds) Introduction to Social Psychology (4th edition), Blackwell Publishing, London.
45. Hinds AL, Woody EZ, Drandic A, Schmidt LA, Van Ameringen M, Coroneos M, Szechtman H (2010) The psychology of potential threat: properties of the security motivation system. Biol Psychol 85(2):331-337 doi:10.1016/j.biopsycho.2010.08.003
46. Hirschberger G, Pyszczynski T, Ein-Dor T (2009) Vulnerability and vigilance: threat awareness and perceived adversary intent moderate the impact of mortality salience on intergroup violence. Pers Soc Psychol Bull 35(5):597-607 doi:10.1177/0146167208331093
47. Horney JA, Macdonald PD, Van Willigen M, Berke PR, Kaufman JS (2010) Individual actual or perceived property flood risk: did it predict evacuation from Hurricane Isabel in North Carolina, 2003? Risk Anal : Off Publ Soc Risk Anal 30(3):501-511 doi:10.1111/j.1539-6924.2009.01341.x
48. Horswill MS, McKenna FP (2004) Drivers' hazard perception ability: Situation awareness on the road. In: Banbury S, Tremblay S (eds) A cognitive approach to situation awareness: Theory and application, Ashgate Publishing, Ltd., Farnham, UK. pp 155-175
49. Houts PS, Lindell MK, T-w H, Cleary PD, Tokuhata G, Flynn CB (1984) Protective Action Decision Model Applied to Evacuation

During the Three Mile Island Crisis. Int J Mass Emerg Disaster 2(1):27-39

50. Huang SK, Lindell MK, Prater CS, Wu HC, Siebeneck LK (2012) Household Evacuation Decision Making in Response to Hurricane Ike. Nat Hazards Rev 13(4):283-296 doi:10.1061/(Asce)Nh.1527-6996.0000074
51. ISO/IEC (2008) Fire safety —Vocabulary. vol ISO-13943. ISO
52. Jönsson A, Andersson J, Nilsson D (2012) A Risk Perception Analysis of Elevator Evacuation in High-Rise Buildings. In: Shields J (ed) Fifth International Symposium on Human Behaviour in Fire, Interscience Publications, Cambridge, UK. pp 398-410
53. Kahneman D (2011) Thinking, fast and slow. Macmillan
54. Kahneman D, Tversky A (1979) Prospect theory: An analysis of decision under risk. Econometrica: J Econ Soc :263–291
55. Kang JE, Lindell MK, Prater CS (2007) Hurricane Evacuation Expectations and Actual Behavior in Hurricane Lili1. J Appl Soc Psychol 37(4):887-903
56. Kasperson RE, Renn O, Slovic P, Brown HS, Emel J, Goble R, Kasperson JX, Ratick S (1988) The Social Amplification of Risk - a Conceptual-Framework. Risk Anal 8(2):177-187 doi:10.1111/j.1539-6924.1988.tb01168.x
57. Kates RW (1971) Natural Hazard in Human Ecological Perspective - Hypotheses and Models. Econ Geogr 47(3):438-451 doi:10.2307/142820
58. Khan KS, Kunz R, Kleijnen J, Antes G (2003) Five steps to conducting a systematic review. J R Soc Med 96(3):118-121
59. Kinateder M, Pauli P, Muller M, Krieger J, Heimbecher F, Ronnau I, Bergerhausen U, Vollmann G, Vogt P, Muhlberger A (2013) Human behaviour in severe tunnel accidents: Effects of information and behavioural training. Transp Res F-Traf 17:20-32 doi:10.1016/j.trf.2012.09.001
60. Kinateder M, Müller M, Jost M, Mühlberger A, Pauli P (2014a) Social influence in a virtual tunnel fire – influence of conflicting information on evacuation behavior. Appl Ergon
61. Kinateder M, Ronchi E, Gromer D, Müller M, Jost M, Nehfischer M, Mühlberger A, Pauli P (2014) Social influence on route choice in a virtual reality tunnel fire. Transport Res F: Traffic Psychol Behav 6:116-125

62. Kobes M, Helsloot I, deVries B, Post JG (2010) Building safety and human behaviour in fire: A literature review. Fire Saf J 45(1):1-11 doi:DOI 10.1016/j.firesaf.2009.08.005

63. Kuligowski ED (2011) Terror defeated: occupant sensemaking, decision-making and protective action in the 2001 World Trade Center Disaster. Ph.D. Dissertation, Ph.D. Dissertation, University of Colorado at Boulder, Boulder, CO, USA

64. Kuligowski E (2013) Predicting Human Behavior During Fires. Fire Technol 49(1):101-120 doi:10.1007/s10694-011-0245-6

65. Kuligowski E, Gwynne SV (2010) The Need for Behavioral Theory in Evacuation Modeling. In: Klingsch WWF, Rogsch C, Schadschneider A, Schreckenberg M (eds) Pedestrian and Evacuation Dynamics 2008. Springer Berlin Heidelberg, pp 721–732. doi:10.1007/978-3-642-04504-2_70

66. Kuligowski ED, Mileti DS (2009) Modeling pre-evacuation delay by occupants in World Trade Center Towers 1 and 2 on September 11, 2001. Fire Saf J 44(4):487-496 doi:10.1016/j.firesaf.2008.10.001

67. Kuligowski ED, Omori H (2014) General Guidance on Emergency Communication Strategies for Buildings - 2nd Edition NIST TN - 1827. US Department of Commerce, National Institute of Standards and Technology

68. Kuligowski E, Peacock R, Hoskins B (2010) A Review of building evacuation models NIST, Fire Research Division. Technical Note 1680 Washington, US

69. Lazarus RS, Folkman S (1984) Stress, appraisal, and coping. Springer Publishing Company, New York.

70. Lerner JS, Gonzalez RM, Small DA, Fischhoff B (2003) Effects of fear and anger on perceived risks of terrorism: a national field experiment. Psychol Sci 14(2):144-150

71. Lichtenstein S, Slovic P, Fischhoff B, Layman M, Combs B (1978) Judged Frequency of Lethal Events. J Exp Psychol Learn Mem Cogn 4(6):551-578 doi:10.1037//0278-7393.4.6.551

72. Lindell MK, Perry RW (1992) Behavioral foundations of community emergency planning. Hemisphere Publishing Corp

73. Lindell MK, Perry RW (2012) The protective action decision model: theoretical modifications and additional evidence. Risk

Anal : Off Publ Soc Risk Anal 32(4):616-632 doi:10.1111/j.1539-6924.2011.01647.x

74. Lindell MK, Whitney DJ (2000) Correlates of household seismic hazard adjustment adoption. Risk Anal : Off Publ Soc Risk Anal 20(1):13-25
75. Lindell MK, Lu J-C, Prater CS (2005) Household decision making and evacuation in response to Hurricane Lili. Nat Hazards Rev 6(4):171-179
76. Lo SM, Huang HC, Wang P, Yuen KK (2006) A game theory based exit selection model for evacuation. Fire Saf J 41(5):364-369 doi:10.1016/j.firesaf.2006.02.003
77. Loewenstein GF, Weber EU, Hsee CK, Welch N (2001) Risk as feelings. Psychol Bull 127(2):267-286 doi:10.1037//0033-2909.127.2.267
78. Martin WE, Martin IM, Kent B (2009) The role of risk perceptions in the risk mitigation process: the case of wildfire in high risk communities. J Environ Manag 91(2):489-498 doi:10.1016/j.jenvman.2009.09.007
79. Mathews A, MacLeod C (1985) Selective processing of threat cues in anxiety states. Behav Res Ther 23(5):563-569
80. Matyas C, Srinivasan S, Cahyanto I, Thapa B, Pennington-Gray L, Villegas J (2011) Risk perception and evacuation decisions of Florida tourists under hurricane threats: a stated preference analysis. Nat Hazards 59(2):871-890 doi:10.1007/s11069-011-9801-0
81. Mbaye S, Kouabenan DR (2013) Effects of the feeling of invulnerability and the feeling of control on motivation to participate in experience-based analysis, by type of risk. Accid Anal Prev 51:310-317 doi:10.1016/j.aap.2012.11.026
82. McConnell NC, Boyce KE (2013) Refuge areas and vertical evacuation of multistorey buildings: the end users' perspectives. Fire Mater. doi:10.1002/fam.2205
83. McConnell NC, Boyce KE, Shields J, Galea ER, Day RC, Hulse LM (2010) The UK 9/11 evacuation study: Analysis of survivors' recognition and response phase in WTC1. Fire Saf J 45(1):21-34 doi:10.1016/j.firesaf.2009.09.001

84. Mccrae RR, Costa PT (1995) Trait Explanations in Personality Psychology. Eur J Personal 9(4):231-252 doi:10.1002/per.2410090402

85. McGee TK, Russell S (2003) "It's just a natural way of life..." an investigation of wildfire preparedness in rural Australia. Global Environ Chang B: Environ Hazards 5(1):1-12

86. McKenna F, Crick J (1991) Experience and expertise in hazard perception. In: Behavioural research in road safety, Nottingham, GB, 26–27 September 1990. vol PA 2038/91

87. McLaughlin AC, Mayhorn CB (2014) Designing effective risk communications for older adults. Saf Sci 61:59-65 doi:10.1016/j.ssci.2012.05.002

88. Miceli R, Sotgiu I, Settanni M (2008) Disaster preparedness and perception of flood risk: A study in an alpine valley in Italy. J Environ Psychol 28(2):164-173 doi:10.1016/j.jenvp.2007.10.006

89. Michalsen A (2003) Risk assessment and perception. Inj Control Saf Promot 10(4):201-204 doi:10.1076/icsp.10.4.201.16782

90. Mileti DS, Sorensen JH (1990) Communication of emergency public warnings: A social science perspective and state-of-the-art assessment. Oak Ridge National Lab, TN (USA).

91. Mongrain S, Standing L (1989) Impairment of cognition, risk-taking, and self-perception by alcohol. Percept Mot Skills 69(1):199-210

92. Nilsson D, Johansson A (2009) Social influence during the initial phase of a fire evacuation—Analysis of evacuation experiments in a cinema theatre. Fire Saf J 44(1):71-79

93. Ogilvie D, Egan M, Hamilton V, Petticrew M (2005) Systematic reviews of health effects of social interventions: 2. Best available evidence: how low should you go? J Epidemiol Community Health 59(10):886-892

94. Ogilvie D, Hamilton V, Egan M, Petticrew M (2005) Systematic reviews of health effects of social interventions: 1. Finding the evidence: how far should you go? J Epidemiol Community Health 59(9):804-808

95. Öhman A (2000) Fear and anxiety: Evolutionary, cognitive, and clinical perspectives. In: Lewis M, Haviland-Jones J (eds) Handbook of emotions, The Guilford Press, New York. pp 573-593

96. Okabe K, Mikami S (1982) A study on the socio-psychological effect of a false warning of the Tokai Earthquake in Japan. In: A Paper presented at the Tenth World Congress of Sociology, Mexico City, Mexico

97. Patterson O, Weil F, Patel K (2010) The Role of Community in Disaster Response: Conceptual Models. Popul Res Policy Rev 29(2):127-141 doi:10.1007/s11113-009-9133-x

98. Perry RW (1979) Evacuation Decision-Making in Natural Disasters. Mass Emerg 4(1):25-38

99. Perry RW, Lindell MK (2004) Communicating Environmental Risk in Multiethnic Communities. Sage Publications, Thousand Oaks, CA.

100. Pradhan AK, Fisher DL, Pollatsek A (2006) Risk perception training for novice drivers - Evaluating duration of effects of training on a driving simulator. Transp Res Rec 1969(1969):58-64

101. Proulx G (1993) A Stress Model for People Facing a Fire. J Environ Psychol 13(2):137-147 doi:10.1016/S0272-4944(05)80146-X

102. Proulx G (1995) Evacuation Time and Movement in Apartment Buildings. Fire Saf J 24(3):229-246 doi:10.1016/0379-7112(95)00023-M

103. Proulx G, et al. (2008) SFPE Handbook of Fire Protection Engineering.DiNenne P, Drysdale D, Beyler C (eds) National Fire Protection Association, Quincey, MA. pp 3-355-353-372

104. Purser DA, Bensilum M (2001) Quantification of behaviour for engineering design standards and escape time calculations. Saf Sci 38(2):157-182 doi:10.1016/S0925-7535(00)00066-7

105. Rando CM, Patel DG, Duvall LE, Martin L (2007) Use of Cautions and Warnings within International Space Station Procedures: When Too Much Information Becomes Risky. Proc Hum Factors Ergon Soc Annu Meet 51(20):1435-1438 doi:10.1177/154193120705102013

106. Rayner S, Cantor R (1987) How Fair Is Safe Enough? The Cultural Approach to Societal Technology Choice1. Risk Anal 7(1):3-9

107. Reneke PA (2013) NIST IR 7914 Evacuation Decision Model. NIST Interagency/Internal Report. US Department of Commerce, National Institute of Standards and Technology, Gaithersburg, MS, USA.

108. Riad JK, Norris FH, Ruback RB (1999) Predicting Evacuation in Two Major Disasters: Risk Perception, Social Influence, and Access to Resources1. J Appl Soc Psychol 29(5):918-934 doi:10.1111/j.1559-1816.1999.tb00132.x
109. Riley D (2014) Mental models in warnings message design: A review and two case studies. Saf Sci 61:11-20 doi:10.1016/j.ssci.2013.07.009
110. Robins RW, Hendin HM, Trzesniewski KH (2001) Measuring global self-esteem: Construct validation of a single-item measure and the Rosenberg self-esteem scale. Personal Soc Psychol Bull 27(2):151-161 doi:10.1177/0146167201272002
111. Rogers RW, Prentice-Dunn S (1997) Protection motivation theory. In: Gochman D (ed) Handbook of Health Behavior: Vol. 1 Determinants of Health Behavior: Personal and Social, Plenum, New York. pp 113-132
112. Ronchi E, Nilsson D (2013) Fire evacuation in high-rise buildings: a review of human behaviour and modelling research. Fire Sci Rev 2(1):7 doi:10.1186/2193-0414-2-7
113. Ryb GE, Dischinger PC, Kufera JA, Read KM (2006) Risk perception and impulsivity: association with risky behaviors and substance abuse disorders. Accid Anal Prev 38(3):567-573 doi:10.1016/j.aap.2005.12.001
114. Sarter NB, Woods DD (1991) Situation awareness: A critical but ill-defined phenomenon. Int J Aviat Psychol 1(1):45-57
115. Schacter D, Gilbert D, Wegner D (2011) Sensation and Perception. Charles Linsmeiser Psychology Worth Publishers p 158 159
116. Sheppard BH, Hartwick J, Warshaw PR (1988) The Theory of Reasoned Action - a Meta-Analysis of Past Research with Recommendations for Modifications and Future-Research. J Consum Res 15(3):325-343 doi:10.1086/209170
117. Sherman MF, Peyrot M, Magda LA, Gershon RRM (2011) Modeling pre-evacuation delay by evacuees in World Trade Center Towers 1 and 2 on September 11, 2001: A revisit using regression analysis. Fire Saf J 46(7):414-424 doi:10.1016/j.firesaf.2011.07.001
118. Shields TJ, Boyce KE, McConnell N (2009) The behaviour and evacuation experiences of WTC 9/11 evacuees with self-designated mobility impairments. Fire Saf J 44(6):881-893 doi:10.1016/j.firesaf.2009.04.004

119. Siebeneck LK, Cova TJ (2012) Spatial and temporal variation in evacuee risk perception throughout the evacuation and return-entry process. Risk Anal : Off Publ Soc Risk Anal 32(9):1468-1480 doi:10.1111/j.1539-6924.2011.01781.x
120. Siegrist M, Cvetkovich G (2000) Perception of hazards: the role of social trust and knowledge. Risk Anal : Off Publ Soc Risk Anal 20(5):713-719
121. Simon HA (1972) Theories of bounded rationality. Decision Organ 1:161-176
122. Simon HA (1991) Bounded Rationality and Organizational Learning. Organ Sci 2(1):125-134 doi:10.1287/Orsc.2.1.125
123. Sjöberg L (2000) Factors in risk perception. Risk Anal 20(1):1-12
124. Sjöberg L, Moen B-E, Rundmo T (2004) Explaining risk perception. c Rotunde publikasjoner Trondheim, Norway.
125. Slovic P (1987) Perception of risk. Science 236(4799):280-285
126. Slovic P (2000) The perception of risk. Earthscan Publications
127. Slovic P (2010a) The feeling of risk: new perspectives on risk perception. Routledge
128. Slovic P (2010) The More Who Die, The Less We Care. In: Slovic P (ed) The Feeling of Risk, Routledge, New York, NY. pp 69-78
129. Slovic P, Weber EU (2002) Perception of Risk Posed by Extreme Events. Paper presented at the Columbia-Wharton/Penn Roundtable on "Risk Management Strategies in an Uncertain World" Palisades, New York
130. Slovic P, Peters E, Finucane ML, Macgregor DG (2005) Affect, risk, and decision making. Health Psychol : Off J Div Health Psychol Am Psychol Assoc 24(4 Suppl):S35-S40 doi:10.1037/0278-6133.24.4.S35
131. Slovic P, Finucane ML, Peters E, MacGregor DG (2007) The affect heuristic. Eur J Oper Res 177(3):1333-1352
132. Smerecnik CM, Mesters I, Candel MJ, De Vries H, De Vries NK (2012) Risk perception and information processing: the development and validation of a questionnaire to assess self-reported information processing. Risk Anal : Off Publ Soc Risk Anal 32(1):54-66 doi:10.1111/j.1539-6924.2011.01651.x

133. Song H, Schwarz N (2009) If It's Difficult to Pronounce, It Must Be Risky Fluency, Familiarity, and Risk Perception. Psychol Sci 20(2):135-138

134. Stankowich T, Blumstein DT (2005) Fear in animals: a meta-analysis and review of risk assessment. Proc Biol Sci R Soc 272(1581):2627-2634 doi:10.1098/rspb.2005.3251

135. Strack F, Deutsch R (2004) Reflective and impulsive determinants of social behavior. Pers Soc Psychol Rev : Off J Soc Pers Soc Psychol Inc 8(3):220-247 doi:10.1207/s15327957pspr0803_1

136. Szechtman H, Woody E (2004) Obsessive-compulsive disorder as a disturbance of security motivation. Psychol Rev 111(1):111-127 doi:10.1037/0033-295X.111.1.111

137. Teigen KH (2005) The proximity heuristic in judgments of accident probabilities. Br J Psychol 96(Pt 4):423-440 doi:10.1348/000712605X47431

138. Thompson CP, Mingay D (1991) Estimating the Frequency of Everyday Events. Appl Cogn Psychol 5(6):497-510 doi:10.1002/acp.2350050605

139. Thompson N, Stradling S, Murphy M, ONeill P (1996) Stress and organizational culture. Br J Soc Work 26(5):647-665

140. Trower P, Gilbert P, Sherling G (1990) Social anxiety, evolution and self-presentation. In: Leitenberg H (ed) Handbook of Social and Evaluation Anxiety. Springer, pp 11–46

141. Trumbo CW, McComas KA (2003) The function of credibility in information processing for risk perception. Risk Anal : Off Publ Soc Risk Anal 23(2):343-353

142. Ulleberg P, Rundmo T (2003) Personality, attitudes and risk perception as predictors of risky driving behaviour among young drivers. Saf Sci 41(5):427-443 doi:10.1016/S0925-7535(01)00077-7

143. Wachinger G, Renn O, Begg C, Kuhlicke C (2012) The risk perception paradox—implications for governance and communication of natural hazards. Risk Anal

144. Wanous JP, Reichers AE, Hudy MJ (1997) Overall job satisfaction: how good are single-item measures? J Appl Psychol 82(2):247

145. Watts J, Hall J, et al. (2008) Introduction to Fire Risk Analysis. In: DiNenne P, Drysdale D, Beyler C (eds) SFPE Handbook of

Fire Protection Engineering, National Fire Protection Association, Quincey, MA. pp 5.1-5.8

146. Weber EU, Blais AR, Betz NE (2002) A domain-specific risk-attitude scale: Measuring risk perceptions and risk behaviors. J Behav Decis Mak 15(4):263 doi:10.1002/Bdm.414

147. Wilson F, Gott M, Ingleton C (2013) Perceived risks around choice and decision making at end-of-life: a literature review. Palliat Med 27(1):38-53 doi:10.1177/0269216311424632

148. Wisner B (2004) At risk: natural hazards, people's vulnerability and disasters. Psychology Press, Oxford, UK

149. Wogalter MS, DeJoy D, Laughery KR (1999) Warnings and risk communication. CRC Press, Boca Raton, FL, USA

150. Woody EZ, Szechtman H (2011) Adaptation to potential threat: the evolution, neurobiology, and psychopathology of the security motivation system. Neurosci Biobehav Rev 35(4):1019-1033 doi:10.1016/j.neubiorev.2010.08.003

151. Woody EZ, Szechtman H (2013) A biological security motivation system for potential threats: are there implications for policy-making? Frontiers in human neuroscience 7

152. Yiend J (2010) The effects of emotion on attention: A review of attentional processing of emotional information. Cogn Emot 24(1):3-47 doi:10.1080/02699930903205698

153. Yuan JP, Fang Z, Wang YC, Lo SM, Wang P (2009) Integrated network approach of evacuation simulation for large complex buildings. Fire Saf J 44(2):266-275 doi:10.1016/j.firesaf.2008.07.004

154. Zheng XP, Cheng YA (2011) Conflict game in evacuation process: A study combining Cellular Automata model. Physica A 390(6):1042-1050 doi:10.1016/j.physa.2010.12.007

155. Zuckerman M, Kuhlman DM (2000) Personality and Risk-Taking: Common Bisocial Factors. J Pers 68(6):999-1029

Chapter 5

Remote Control System of Disaster Response Robot with Passive Sub-Crawlers Considering Falling Down Avoidance

Soichiro Suzuki[1], Satoshi Hasegawa[2], and Masayuki Okugawa[3]

[1]Graduate School of Engineering, Aichi Institute of Technology, 1247 Yachigusa, Yakusa, Toyota, Aichi 4700369, Japan

[2]Daiichi System Engineering, 4F Green Bldg, 1-16-20 Nishiki Naka Nagoya, Aichi 460003, Japan

[3]Aichi Institute of Technology, 1247 Yachigusa, Yakusa, Toyota, Aichi, 4700369, Japan

ABSTRACT

This paper describes a remote control system for a crawler-type mobile robot with passive sub-crawlers. This system has a great advantage because it utilizes an essential compliant mechanism that allows the angle of the sub-crawlers to be adapted to the shape of the road surface. Its operation is extremely simple, and it is only necessary to control the movement direction and driving speed in comparison with the case of controlling active sub-crawlers. However, a robot with passive sub-crawlers cannot recover from a situation in which it is stuck. The operator must select a traversable route for unknown rough terrain using only the information obtained from camera images and some sensor data from the robot. In this study, a remote control system for a crawler robot with passive sub-crawlers was developed based on a warning system. This system evaluates the currently selected route by calculating the stabilization for the robot when falling down in the roll and pitch directions. Experimental results obtained using a prototype crawler robot with passive sub-crawlers demonstrated the effectiveness of the proposed system.

BACKGROUND

Remote-controlled mobile robots are useful for searching around and inside buildings that have collapsed in a disaster. In a disaster area, to avoid the risk of secondary disasters, it is preferable to immediately deploy remote-controlled mobile robots instead of waiting for first responders such as firefighters. Disaster response robots should have high mobility on rough terrain. The movement mechanism of a crawler-type robot with sub-crawlers has been applied to many disaster response robots in order to realize high mobility [1]. However, its use complicates robot operation because the operator must actively control each sub-crawler by estimating the attitude or state of the robot based on camera images and sensor information. Therefore, because of the multiple degrees of freedom involved, operators must be well trained to achieve high mobility by using remote control. A semi-autonomous control system for crawler robots has been studied to realize both high mobility and simple operation [2],[3]. This study focused on a crawler robot with passive sub-crawlers to realize simple operation with multiple degrees of freedom. Passive sub-crawlers can adapt to unknown rough terrain. We developed a crawler-type mobile robot for conducting search operations around a disaster area. Passive sub-crawlers called "Scott[a] I" were adopted for this robot. In a crawler robot with passive sub-crawlers, it is only necessary to control the movement direction and driving speed, unlike in the case of controlling active sub-crawlers. However, posture control is impossible in passive sub-crawlers because the operator cannot actively control each sub-crawler, and a robot with passive sub-crawlers cannot recover from a situation in which it is stuck. A robot with passive sub-crawlers is extremely simple, but it is essential for the operator to possess considerable skill when selecting a route. The operator must select a traversable route for unknown rough terrain using only the information obtained from camera images and sensor data from the robot.

Abbreviation for "Scouting Crawler Robot Technology".

In this paper, we propose a warning system, not a semi-autonomous control system, for a remote-controlled disaster response robot, because the operator should have the authority to make the final decision in order to realize flexible operation at all times. By evaluating the stability margin, a remote control system for a crawler robot with passive sub-crawlers is proposed that adopts the warning system with a dynamic threshold. The current route selected by the operator is evaluated by calculating the stabilization for the robot while falling down in the roll and pitch directions. An operator can prevent the robot from falling down based on the sound and signals from the warning system when the evaluation value for the stability margin is less than the threshold. The proposed system will be able to prompt the operator to select another route. A method of compensating for the disadvantages of a crawler robot with passive sub-crawlers will be proposed in this paper. This paper describes a method for evaluating the potential for falling down by using a predictive falling down margin time based on the normalized energy (NE) stability margin. The results of experiments demonstrated the effectiveness of the proposed system.

DISASTER RESPONSE ROBOT "SCOTT I"

Figure 1(a) shows an overview of the "Scott I" crawler-type mobile robot. Table 1 lists the robot's specifications. The driving mechanism of the main crawler includes a drive motor and four sub-crawlers (front × 2, rear × 2). Figure 1(b) shows a cross-sectional view of the sub-crawlers' joint part. Each rotary joint of the sub-crawlers can rotate freely and has passivity. The rotation of a sub-crawler's belt is transmitted from the main crawler. However, the rotations of the right and left sub-crawlers are not independent.

(a) Overview of "Scott I"

(b) Mechanism of passive sub-crawlers

Figure 1: Image of crawler-type mobile robot with passive sub-crawlers: Scott I.

Table 1: Specifications of Scott 1

Dim. of the robot [mm]	Length × Width × Height
Max.	720 × 390 × 145
Min.	370 × 290 × 100
Timing pulley dia. [mm]	
Large	145
Small	100
Belt width [mm]	
Main crawler	50.8
Sub-crawlers	25.4
Weight [kg]	20.0
Max. speed [m/s]	0.35

Suzuki et al.

Suzuki et al. ROBOMECH Journal 2014 1:20, doi:10.1186/s40648-014-0020-9

The operation of a robot with active sub-crawlers is complicated because the operator should control not only the main body but also each sub-crawler. A semi- autonomous control system has been considered to overcome the disadvantage of a crawler-type robot with active sub-crawlers. We do not intend to deny the approach for a semi-autonomous control system. A crawler-type robot with passive sub-crawlers is known to have a great advantage because it has an essential compliant mechanism that allows the angle of the sub-crawlers to be adapted to the shape of the road surface [4],[5]. The operator only needs to control the movement direction and driving speed. The effectiveness of passive sub-crawlers has been confirmed by simulations and experiments, and the mobility performance of Scott I was appreciated in RoboCup Japan Open Rescue Robot League 2013 and in the evaluation field at IRS (International Rescue System Institute in Japan). On the other hand, it is necessary to restrict the rotational angle of a sub-crawler to avoid a singular configuration when moving over a step, and numerical simulations and experiments show that the restricting angle differs with the step height and road shape. Scott I can change the restriction angle of the sub-crawlers by adjusting the length of the belt and stopper member. However, the restriction angle cannot be changed in real time. Therefore, a subject for future work will be a control method to set a suitable restriction angle based on the step height and road-surface shape.

Principle of Traversing on Simple Step for Crawler-Type Robot with Passive Sub-Crawlers

The principle of traversing a simple step for a crawler-type robot with passive sub-crawlers is described. If the diameters of the pulleys for the sub-crawlers are the same, only a forward driving force is generated. Therefore, no upward power is generated, and the center of gravity of the main body of the robot cannot be moved above the obstacle. On the other hand, after the tip of a sub-crawler makes contact with an obstacle, a vertical force acts, and the sub-crawler is rotated upward by the larger pulley of the contact side. The robot will begin to traverse the step by increasing the driving force to a value greater than the weight of the robot itself.

Figure 2 shows the sequence when a crawler-type robot with passive sub-crawlers traverses a simple step. This sequence was confirmed by the results of simulations and experiments. The traversing sequence for a simple step is explained as follows. The height of a simple step is defined as greater than the radius of the large forward timing pulley, and less than the height of the center of gravity on the perpendicular robot.

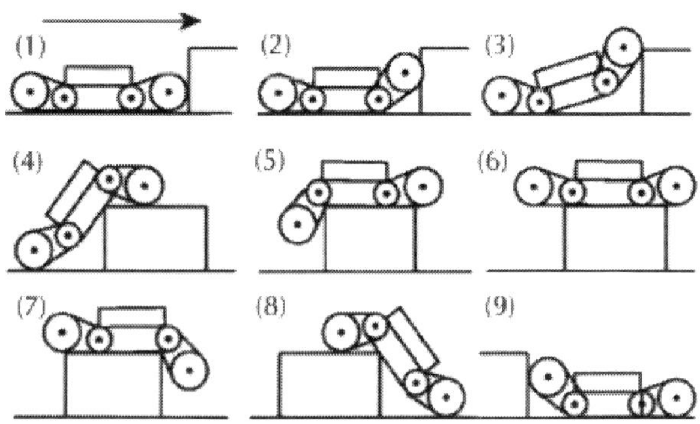

Figure 2: Sequence for traversing simple step for crawler-type mobile robot with passive sub-crawlers.

- The force acts vertically.
- The front sub-crawlers press against the surface of the step and rotate upward.
- The attitude angle of the main crawler increases.
- The height of the center of gravity is higher than the objective step and moves forward from the rotational center of the rear sub-crawlers' joint.
- The front sub-crawlers and main body move over the step.
- The entire robot moves over the step.
- The front sub-crawlers rotate downward.
- The front sub-crawlers make contact with the ground.
- The rear sub-crawlers make contact with the ground, and the robot finishes moving over the step.

WARNING SYSTEM

Necessity of Warning System

As mentioned in the previous section, the operation is simple because a crawler robot with passive sub-crawlers can traverse rough terrain using only instructions regarding the moving direction and driving speed as long as sufficient ground pressure is applied, and the rotational angle of the sub-crawlers is appropriately restricted. However, posture control using the sub-crawlers is impossible because the operator cannot control the sub-crawlers individually. Therefore, the operator predicts the traversal of an obstacle or the state of the robot given sufficient ground pressure, and they must select an appropriate route using only the camera image and sensor information obtained from the robot (Figure 3). One problem lies in the fact that the operator must master skills for route selection in order to prevent the robot from falling down.

Figure 3: Which route should the operator select?

This paper proposes a warning system to assist the operator with route selection. This system can indicate and evaluate the stability margin for the potential of falling down by using the angle information for the robot's body in the pitch and roll directions. To prevent the robot from falling down, the proposed system can provide a warning to the operator when the stability index is below the threshold value by quantitatively evaluating the current route. A flowchart for the proposed system is shown in Figure 4. We evaluate the stability of the robot on the basis of the NE stability margin. This system has the ability to support operations in a narrow space without depending on images from an overlooking camera.

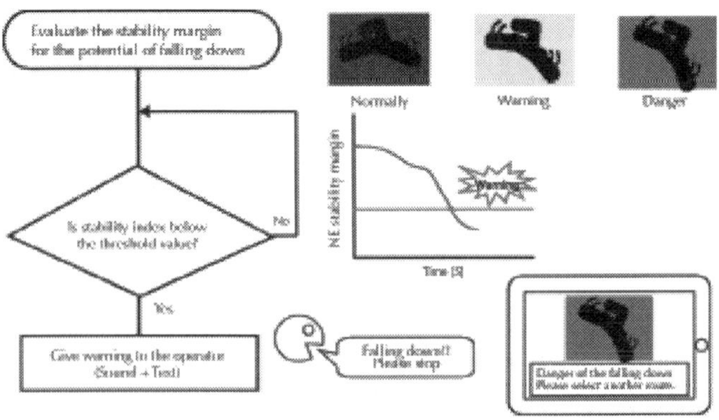

Figure 4: Flowchart for warning system.

Derivation of Normalization Energy Stability Margin

The NE stability margin $E_n(t)$ is a stability index for a robot. It is based on the difference between the maximum height h_{max} and the current height $h(t)$ [mm] for the center of gravity when the robot body rotates around the pitch and roll axes.

$$E_n(t) = h_{max} - h(t) \tag{1}$$

In this study, $h_p(t)$ and $h_r(t)$ were defined as the $h(t)$ values in the pitch and the roll directions, respectively. The robot's center of gravity $h(t)$ was calculated using the body angles in the pitch$\theta b(t)$ [rad], and roll $\theta w(t)$ [rad] directions, along with the relative angle of the sub-crawlers $\theta_f(t)$ to the main crawler. $\theta_b(t)$ was measured using a gyroscope, and $\theta_f(t)$ and $\theta_r(t)$ were measured using a potentiometer. Figure 5(a) shows the modeling of the robot and parameters for the pitch direction.

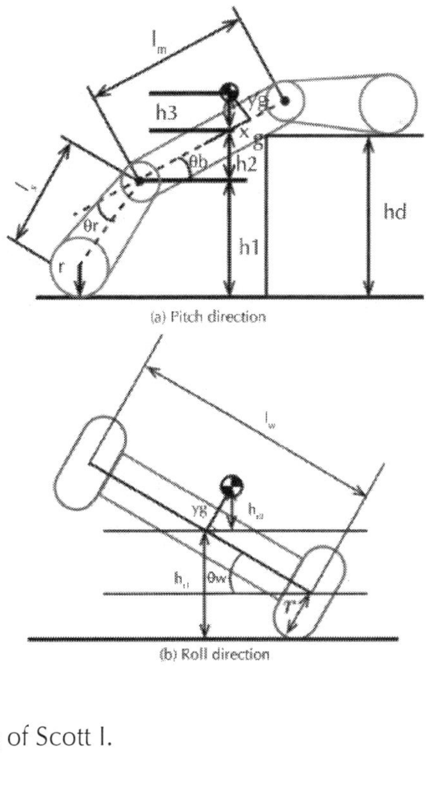

Figure 5: Modeling of Scott I.

The length of the main crawler is l_m [mm], length of the sub-crawlers is l_s [mm], and radius of the pulley is r [mm].

In the initial state, y_g [mm] is the vertical coordinate of the center of gravity from the road surface, and x_g [mm] is the horizontal coordinate of the center of gravity from the center of the main body. When moving up stairs, $h_{p1}(t)$ is expressed by equation (2). When moving down stairs,$hp1'(t)$ is expressed by equation (3).

$$h_{p1}(t) = l_s \sin(\theta_r(t) + \theta_b(t)) + r \tag{2}$$

$$h'_{p1}(t) = l_s \sin\left(\theta_f(t) + \theta_b(t)\right) + r \tag{3}$$

$$h_{p2}(t) = \left(\frac{1}{2}l_m + x_g - y_g \tan\theta_b(t)\right)\sin\theta_b(t) \tag{4}$$

$$h_{p3}(t) = \frac{y_g}{\cos\theta_b(t)} \tag{5}$$

$$h_p(t) = h_{p1}(t) + h_{p2}(t) + h_{p3}(t) \tag{6}$$

$$h'_p(t) = h'_{p1}(t) + h_{p2}(t) + h_{p3}(t) \tag{7}$$

The coordinate of the center of gravity for the roll direction is calculated by measuring the body angle in the roll direction ($-\pi/2<\theta_w(t)<\pi/2$) using the gyroscope. Figure 5(b) shows the modeling of the robot and parameters in the roll direction.

$$h_{r1}(t) = \frac{1}{2}l_w \sin|\theta_w(t)| + r\cos\theta_w(t) \tag{8}$$

$$h_{r2}(t) = y_g \cos\theta_w(t) \tag{9}$$

$$h_r(t) = h_{r1}(t) + h_{r2}(t) \tag{10}$$

Verification of Estimated Center of Gravity

The validity of equations (6), (7), and (10) is confirmed by comparing the estimated value with the experimental one. The estimated value is effective in the range of expression (11) provided that the crawler is always in contact with the ground. Where, h_d is defined it as the height of the step.

$$r < h_d < x_g + \frac{1}{2} l_m + l_s + r \tag{11}$$

The range of the rotational angle for the sub-crawlers is limited to values between −40 deg and 30 deg in order to avoid the singular configuration of the front sub-crawlers. It is not necessary to consider the influence of the front sub-crawlers, because the rotation range is narrow. Figure 6shows the experimental and estimated values. Figure 6(a) and (b) shows that the estimated value agrees very well with the experimental one. The error observed from 4—6 sec in Figure 6(a) occurs because the sub-crawlers do not make contact with the ground. The error observed from 9—12 sec is the spike noise included in the sensor data for the angular angular velocity measured using a gyroscope when the robot moves over the step. An experiment in the roll direction was performed for a robot traversing a slope on one side of the crawler. The experimental and estimated values are shown in Figure 6(b). The roll direction of the center of gravity was exactly estimated for the sub-crawlers that were always in contact with the ground. The experimental values of the center of gravity were calculated using the coordinate obtained from image processing. The marker was attached in order to trace the center of gravity for the robot.

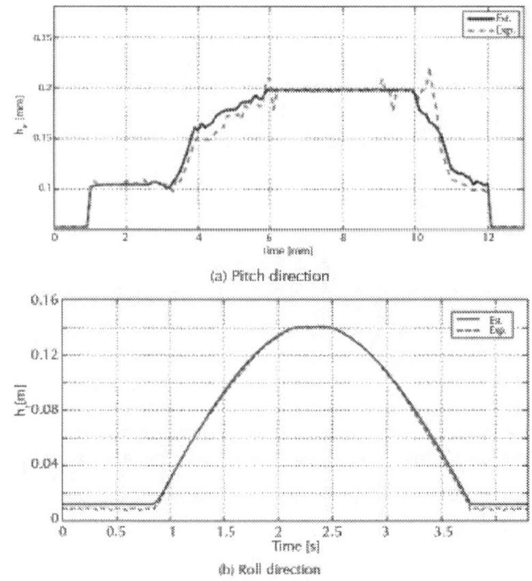

Figure 6: Comparison of estimated and experimental values.

Examination of Threshold for Warning System

The timing of a warning is generated using the threshold of the NE stability margin $E_n(t)$. It is possible to prevent the robot from falling down if the operator takes appropriate action in response to the warning.

In this paper, a predictive falling down margin time *Tm* is adopted to warn the operator that the robot will fall down. T_m is calculated by using the time derivative for the NE stability margin $E_n(t)$, as shown in Figure 7. When $t = \bar{t}$, a gradient $\alpha(\bar{t})$ is defined as $dE_n(\bar{t})/dt$. The equation of a tangential line for *En*(*t*) at $t = \bar{t}$ is derived using equation (12).

$$\overline{E_n}(t) = \alpha(t)t + E_n(t) \quad (12)$$

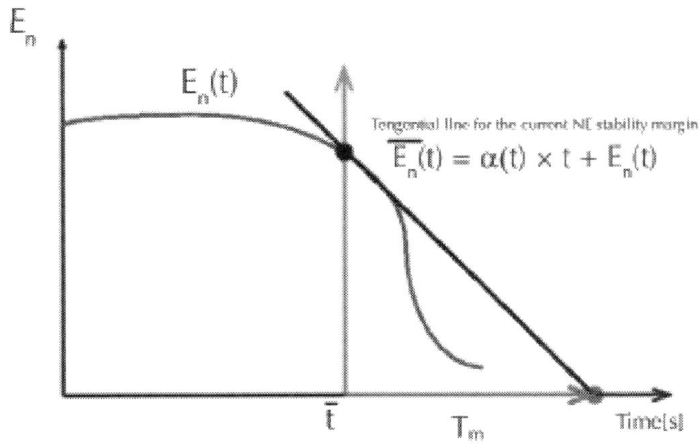

Figure 7: Image of predictive falling down margin time.

Tm is defined as a time before $\bar{E}_n = 0$ from $\bar{t}$ using equation (13).

$$T_m = -\frac{E_n(\bar{t})}{\alpha(\bar{t})} \tag{13}$$

T_r is defined as the reaction time of the operator and T_e is defined as the reflection time for the operator's control. The timing to give the warning should be decided in consideration of T_m and the reaction time of the operator T_r. The reaction time is estimated as T_r=0.4 sec based on reference [6]. The reflection time is estimated to be T_e=0.1 sec in consideration of the time delay for system execution. The threshold time is determined to be T_m=0.5 sec in this paper. In addition, a notice will be displayed 0.4 sec before the warning. The attention time is assigned while $0.5<T_m<1.0$.

Method for Operation Support

As shown in the Figure 8, the robot body posture for the pitch and roll directions and the stability will be displayed on the operation screen to determine the appropriateness of the route selection. The stability of the operation is displayed on the monitor in terms of the rate of $E_n(t)$ for the initial posture of the robot and the present one. Depending on the deterioration of the stability margin, the background color of the display is sequentially changed from blue, to yellow, and finally to red. If the NE stability margin is lower than the threshold, the level of danger will be presented as a warning by voice and a signal because of the risk of the robot falling down.

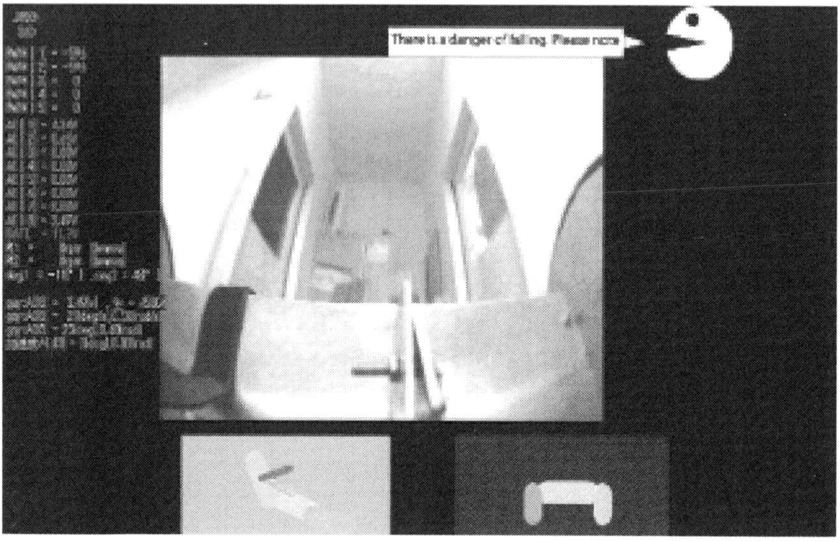

Figure 8: Display method for operation support information.

In the case of a robot with active sub-crawlers, the operator must appropriately control each sub-crawler's angle in every sequence. A robot may fall down if the operator lacks adequate skill even if it is a traversable step. In addition, the difference between the traveling sequences of active and passive sub-crawlers lies in the manner in which the rear sub-crawlers are utilized. Passive sub-crawlers can only rotate using gravity; consequently, no downward power to lift the robot body is generated. Adopting passive sub-crawlers has many advantages. It can reduce the number of high-torque motors, reduce the robot weight, and save energy. The relationship between the traversing condition and the sub-crawlers' restricting angle is reported in another paper.

METHODS

Experimental Operation Verification Using Scott I

We confirmed the process of the proposed system using an experimental verification under a real environment. In the experiment, we observed the progress of the operator response to the warning provided by the proposed system. In the experimental condition for the pitch direction, the target obstacle was a simple step (height: 0.4 m, depth: 0.3 m, width: 0.8 m) with an impossible height that was to be traversed. In the experimental condition for the roll direction, the target obstacle was a slope with a 30 deg angle of inclination. The robot traversed this slope only with the left side of the crawler, as shown in Figure 9(b). The operator screen and the posture of Scott I during this experiment are shown in Figure 9.

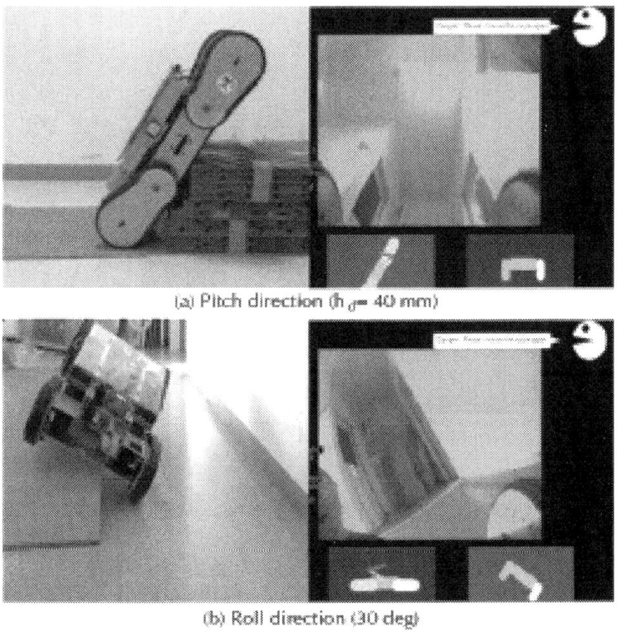

Figure 9: Experimental scene and operator screen.

The time history of the NE stability margin $E_n(t)$ and predictive falling down margin time $T_m(t)$ are shown in Figure 10. The experimental results demonstrate the effectiveness of the proposed system, as shown in Figure 10. We confirmed that the operator could react appropriately to prevent the robot from falling down within approximately 5.0 sec of the warning. However, the robot fell down when the operator did not react at the warning because a strict threshold was assumed. In addition, we could not confirm the effectiveness of the attention time because of the short setting time. We would like to consider these in our future work.

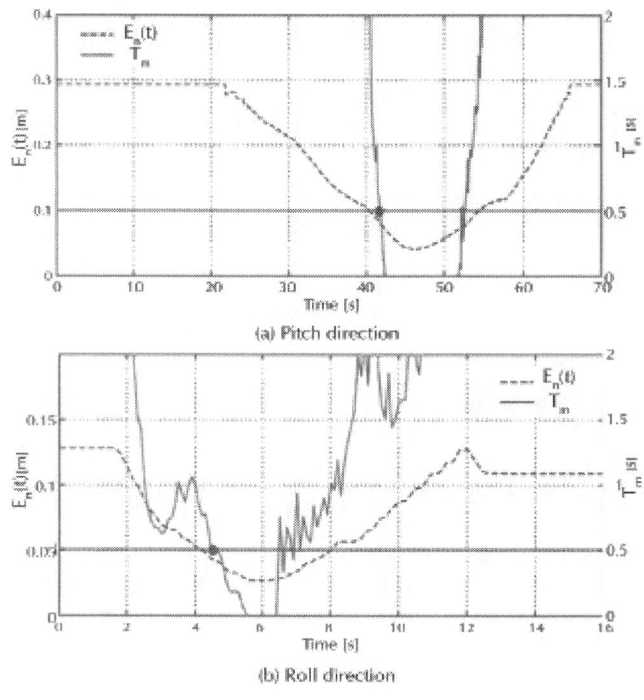

Figure 10: Experimental operation verification of proposed system.

Experimental Evaluation and Consideration

The experimentation for the evaluation of the proposed system was performed with 10 test subjects. We chose expert operators for half the subjects and beginner operators for the rest. On the other hand, the beginner operators were defined as inexperienced operators. The objective of the experiment was to evaluate the reaction of each subject in the operating process for traversing a simple step or slope. The experiment was performed using three cases: Case 1 Using only the images from the front camera (shown in Figure 11(a)) Case 2 Using only the images from the overlooking camera mounted on the manipulator (shown in Figure 11(b)) Case 3 Using the proposed system.

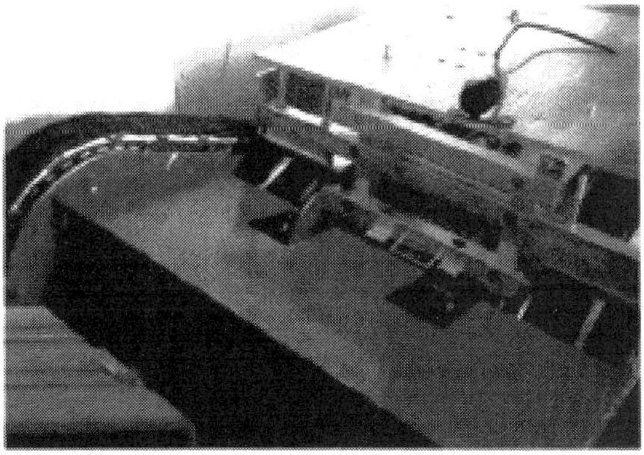

(a) Front camera mounted on main body

(b) Overlooking camera mounted on manipulator

Figure 11: Image of the cameras mounted on Scott.

Each subject performed the operation nine times for these experimental cases.

Equation (14) defines the height of the objective traversing step $\bar{h}$, where h_e [m] is the maximum height of the step that the robot can traverse and ε_h is the adjustable height of the step. The step height was classified three conditions ($\bar{h} = 180, 230, 280$ mm), and we chose randomly it for each test subjects. Each step height condition was chosen to become the same number of experiments.

$$\bar{h} = h_e + \varepsilon_h \tag{14}$$

In the case where the step to be traversed had an untraversable height ($\bar{h} > h_e$)we defined a misjudgment as the situation in which the robot fell down even though the subject could traverse the objective step and the robot could continue to move forward. We also defined a misjudgment as the situation in which the subject operated the robot backward upon understanding that the step to be traversed had an untraversable height.

The experimental results were evaluated by the misjudgment rate R_w based on equation (15), where *Ex* is the number of times the experiment was conducted and M_j is the number of misjudgments.

$$R_w = \frac{M_j}{E_x} \tag{15}$$

However, the same number of experiments was performed for each step height condition. The experiments were only performed in the pitch direction at a controllable constant speed (350 mm/sec) for a sampling period of 0.1 sec.

Figure 12 shows the experimental results for the misjudgment rate. The results for the beginner and expert operators are shown in Figure 12(a) and (b), respectively. The graphs show the misjudgment rates on the horizontal axes, and the step height conditions on a vertical axes. The values of the standard deviations for every each step height show in Table 2. The results and discussion are given below based on observing the performances of the operators and discussing these after the experimentation.

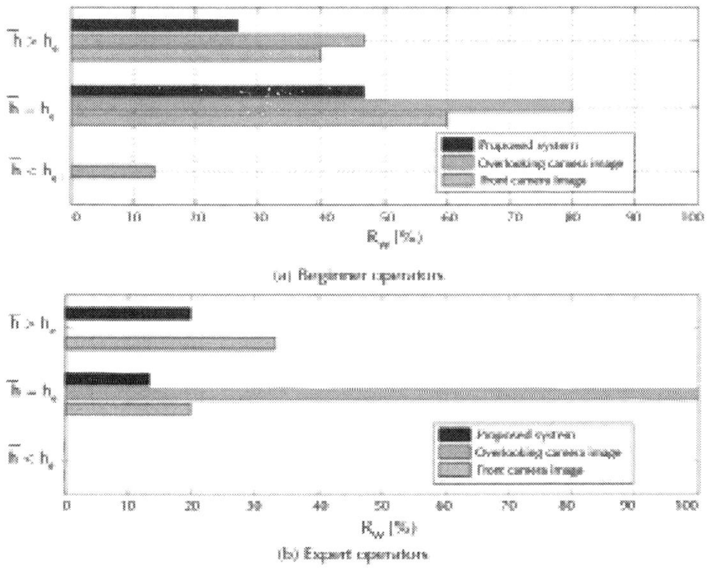

Figure 12: Experimental results for misjudgment rate.

Table 2: Standard deviations for misjudgment rate

Step height		Beginner	Expert
	Case 1	38.9	29.8
$\bar{h} > h_e$	Case 2	26.6	0
	Case 3	24.9	26.7
	Case 1	24.9	26.7
$\bar{h} = h_e$	Case 2	26.6	0
	Case 3	45.2	26.7
	Case 1	0	0
$\bar{h} < h_e$	Case 2	26.7	0
	Case 3	0	0

Suzuki et al.

Suzuki et al. ROBOMECH Journal 2014 1:20, doi: 10.1186/s40648-014-0020-9

RESULTS AND DISCUSSION

First, the results for the beginners are shown in Figure 12(a). In the case of $\bar{h} > h_e$, the appropriate judgment is "move backward" because of the untraversable step height. When using only the images from the front camera (Case 1) or those from the overlooking camera (Case 2), the value of *Rw* was high. From these results, we confirmed that when only using camera images, it is difficult for operators to respond appropriately to prevent the robot from falling down. The obstacle moves out of the camera frame whenever a robot traverses the step because the attitude angle of the robot increases. Therefore, the judgment depends on the experience of the operator. In the case of $\bar{h} > h_e$, we frequently observed the tendency of the cautious judgment to "move backward". As a result, the misjudgment rate in the case of $\bar{h} > h_e$ was smaller than the case of $\bar{h} = h_e$.

On the other hand, when using the warning system (Case 3), the value of *Rw* was the smallest for the untraversable step height $\bar{h} > h_e$. The operators could appropriately control the robot to prevent it from falling down.

Second, in the case of $\bar{h} = h_e$, the value of *Rw* was high under all of the experimental conditions. The "move backward" choice was considered a misjudgment because of the traversable step height. The reasons for Case 1 and Case 2 were similar in the case of $\bar{h} > h_e$. Hence, rash operations and falling down were frequently observed. The reasons for Case 3 was different. The judgment of almost all of the subjects was to "move backward," particularly when using the proposed system. By adopting the proposed warning system, the cautious judgments increased, but rash judgments decreased. The proposed system attached importance to the capability to avoid falling down. When observing the experiments for $\bar{h} = h_e$, it seemed that some subjects were confused because the driving speed of the robot had to be controlled in the case of the limit of the traversable step $\bar{h} = h_e$. This was confirmed by the standard deviation results ($\bar{h} = h_e$), Case 3, Beginner) as listed in Table 2. The driving speed will need be considered in future work.

Finally, for the traversable step height $\bar{h} < h_e$, because the operators only had to select the simple command "Forward" in all cases, the robot was able to traverse the step. The misjudgments of the subjects were almost always "move backward" when using only images from the overlooking camera (Case 2). This result shows that increasing the information to the operator will result in more cautious judgments.

Figure 12(b) shows the experimental results for the misjudgment rate of the expert operators. This rate was smaller than that for the beginners overall. Furthermore, the value of the standard deviation is also small in the case of an expert. We confirmed that the expert operators could perform an appropriate judgments. In the case of $\bar{h} < h_e$, when using only a images from the overlooking camera (Case 2), all of the subjects avoided falling down. The subjects chose the more cautious judgment based on the experience gained in the previous experiments because Case 2 was the last to be performed. In the case of the warning system, we observed that a relatively inexperienced operator let the robot fall down.

In the case of $\bar{h} = h_e$, all of the subjects made wrong judgments for Case 2. The reason was similar to that for the beginner operators. Hence, for Case 1, the misjudgment rate for the expert operators was smaller than the case of $\bar{h} < h_e$ because the obstacle did not frequently move out of the camera frame. When using the proposed system, because it was possible to adjust the movement speed, we confirmed that the high success rate was achieved by repeated stops and starts.

CONCLUSIONS

This paper describes a warning system for a crawler robot with passive sub-crawlers. The proposed support system for the operator could display an evaluation index for the selected route based on the NE stability margin of the robot. Experiments were performed by using Scott I to verify the proposed system. We compared the misjudgment rates between the beginner and expert operators under different experimental conditions. The experimental results demonstrated the effectiveness of the proposed system in all cases. We confirmed that the proposed system was able to stimulate a cautious judgment. Based on the experimental results, we found that it was necessary to provide suitable support based on the skill of the operator. For expert operators, it would be best for the system to provide only assistance. On the other hand, because of their lack of skill, beginner operators cannot react quickly enough to prevent the robot from falling down. Hence, in order to support a beginner operator, we think that it would be best for the system to interrupt their actions, such as through a speed-controlled system. In future work, an experiment will be performed in a real environment by using an improved system to evaluate the effectiveness of the proposed method for expert or beginner operators.

AUTHORS' CONTRIBUTIONS

SH devised the basic concept of the overall system, technically constructed the system and drafted the manuscript. SS carried out the experiments and helped to draft the manuscript. MO suggested the constitution of the article and helped with the editing of the manuscript. All authors read and approved the final manuscript.

ACKNOWLEDGEMENTS

The authors would like to thank Sanritz Automation Co. Ltd. for the development of the computer, "TPIP". This work was supported by a JSPS KAKENHI Grant-in-Aid for Scientific Research (B), No. 23310116.

REFERENCES

1. Koyanagi E (2010) Rescue robot with active sub crawlers. J Robot Soc Japan 28(2):147-150

2. Ohno K, Morimura S, Tadokoro S, Koyanagi E, Yoshida T (2007) Semi-autonomous control system of rescue crawler robot having flippers for getting over unknown-steps In: Proceedings of the IEEE/RSJ international conference on intelligent robots and systems 2007, 3012–3018.
3. Okada Y, Nagatani K, Yoshida K, Yoshida T, Koyanagi E (2010) Shared autonomy system for tracked vehicles to traverse rough terrain based on continuous three-dimensional terrain scanning In: Proceedings of the IEEE/RSJ international conference on intelligent robots and systems 2010, 357–362.
4. Yuuki K, Ikarashi T, Oogane K (2010) Validity of passivity mechanism with rescue robots In: Proceedings of annual conference of robotics society of Japan 2010, 3–14.
5. Kang S, Lee W, Kim M, Shin K (2005) Robhaz-rescue: Rough-terrain negotiable teleoperated mobile robot for rescue mission In: Proceedings of the IEEE/RSJ international conference on intelligent robots and systems 2010, 105–110.
6. Teichner WH, Krebs MJ (1974) Laws of visual choice reaction time. Psychol Rev 81(1):75-98

Chapter 6

Measuring Household Vulnerability to Climate-Induced stresses in Pastoral Rangelands of Kenya: Implications for Resilience Programming

Francis EO Opiyo[1, 2], Oliver V Wasonga[1, 3], and Moses M Nyangito[1]

[1]Department of Land Resources Management and Agricultural Technology, University of Nairobi, Kabete, Nairobi 00625, Kenya

[2]Dryland Development Centre, United Nations Development Programme (UNDP), United Nations Office in Nairobi, Nairobi 00100, Gigiri, Kenya

[3]German Institute for Tropical and Subtropical Agriculture (DITSL), Steinstraße 19, Witzenhausen 37213, Germany

ABSTRACT

This study uses statistical and econometric tools to measure households' vulnerability in pastoral rangelands of Kenya. It considered 27 socio-economic and biophysical indicators obtained from 302 households' in-depth interviews to reflect climate vulnerability components: adaptive capacity, exposure and sensitivity. The theoretical framework used combines exposure and sensitivity to produce potential impact, which was then compared with adaptive capacity in order to generate an overall measure of vulnerability. Principal component analysis (PCA) was used to develop weights for different indicators and produce a household vulnerability index (HVI) so as to classify households according to their level of vulnerability. In order to understand the determinants of vulnerability to climate-induced stresses, an ordered probit model was employed with predictor variables. The results show that 27% of households were highly vulnerable, 44% were moderately vulnerable and 29% of households were less vulnerable to climate-induced stresses. Factor estimates of the probit model further revealed that the main determinants of pastoral vulnerability are sex of household head, age of household head, number of dependents, marital status, social linkages, access to extension services and early warning information, complementary source of income, herd size and diversity, herd structure, herd mobility, distance to markets, employment status, coping strategies and access to credit. Therefore, policies that address these determinants of vulnerability with emphasis on women's empowerment, education and income diversifications are likely to enhance resilience of pastoral households.

BACKGROUND

Vulnerability, commonly defined as the propensity or predisposition to be adversely affected, has been studied as a composite of adaptive capacity, sensitivity and exposure to hazards (Adger and Kelly [1999]; Kelly and Adger [2000]; McCarthy et al. [2001]; Intergovernmental Panel on Climate Change IPCC [2001]; Adger [2006]; Füssel [2007]; Paavola [2008]; Yuga et al. [2010]). Adaptive capacity is the ability of people to cope with or adjust to the changing context and is explained by socio-economic indicators. Sensitivity is the ability of a system to be affected, and exposure is the incidences of events (Kasperson et al. [1995]; Adger [2006]; Paavola [2008]). Vulnerability is thus comprised of risks or a chain of risky events that households confront in pursuit of their livelihoods, the sensitivity of livelihood to these risks, the response or options that households have for managing these risks and finally the outcomes that describe the loss in wellbeing (Turner et al.[2003]).

On the other hand, resilience is seen as the ability to self-organize, learn and adapt to risk hazards (Carpenter et al. [2001]; Turner et al. [2003]). The United Nations International Strategy for Disaster Reduction (United Nations International Strategy for Disaster Reduction-UNISDR[2009]) defines resilience as the transformative process of a household or community exposed to hazards to resist, absorb, accommodate and recover from hazards in a timely and efficient manner. Turner et al. ([2003]) and Gallopın ([2006]) define hazards as threats to a system and comprised of stressors. For simplicity, the term stress is used in this study to denote a continuous or slowly increasing pressure (in this case drought), commonly within the range of normal variability. However, over the last decades, most of the scientific literature and discourses on vulnerability have concentrated on contributing to theoretical insights or measurements at the regional or national scale, with selected indicators for each region, and identifying resilience-building strategies that have implication for national and regional planning (Brooks et al. [2005]; Füssel [2007]; Hinkel [2011]). Yet, micro-level vulnerability analysis is an essential prerequisite for local-level planning and prioritization of resilience planning and strategies especially among the natural resource-dependent communities at risk to projected climate variability and changes (Callaway [2004]; Fraser et al. [2011]).While there is no superior scale of climate vulnerability analysis, recent studies by Yuga et al. ([2010]) and Marshall et al. ([2014]) have confirmed that micro-level analyses have hitherto largely been overlooked in favour of ecosystem-scale studies of biophysical vulnerability.

As observed by Deressa et al. ([2008]), climate change vulnerability analysis ranges from local or household level (Adger [1999]) to the global level (Brooks et al. [2005]; Intergovernmental Panel on Climate Change IPCC [2014]). However, the choice of scale is dictated by the objectives, methodologies and data availability. This present study focused on household-level vulnerability analysis. After all, it is by understanding, planning for and adapting to a changing climate that individual households can take advantage of opportunities to reduce risks associated with climate-induced stresses (Madu [2012]). Similarly, Klein et al. ([2007]) observed that vulnerability analysis to climate change is needed at the level that would enable policy makers to tackle climate change challenges with the precision that is necessary, particularly in the arid and semi-arid regions of Africa. Previous studies by Brooks et al. ([2005]) and Intergovernmental Panel on Climate Change IPCC ([2012]) concluded that the majority of households are particularly vulnerable to the impacts of climate change in the arid and semi-arid regions. Although the causes of such vulnerability are multi-dimensional, they are primarily due to widespread poverty, food insecurity, recurrent droughts, land degradation, inequitable land distribution and overdependence on rain-fed agriculture (Notenbaert et al. [2013]; Lo´pez-Carr et al. [2014]). Some of the other common factors also postulated to determine adaptive capacity, and therefore influence vulnerability at the household level, include access to resources, markets and infrastructure; household structure, gender, education and age; farm size; income and income diversity; access to community-based organizations, information, credit facilities, savings and loans; and health status among others. Stringer et al. ([2009]) argued that though all households in a community are exposed to risks associated with climate change and could potentially be rendered vulnerable, the poorer households are the most at risk of adverse impacts of climate variability and change. That notwithstanding, the proportion of households vulnerable to extreme weather events is perceived to be increasing in arid environments of eastern Africa, especially in Kenya, and there is an uncertain degree to which the population or the system is becoming susceptible and unable to cope with hazards and stresses, including the effects of climate change.

In Kenya, there is a consensus that projected climate change will worsen food security, mainly through increased extremes and temporal or spatial shifts (Eriksen and O'Brien [2007]; Herrero et al. [2010]; Sherwood [2013]). In fact, there is considerable potential impact of these global drivers of change on the production systems and resource-poor households who depend on them. Studies by Kabubo-Mariara ([2009]) and Silvestri et al. ([2012]) revealed that extreme weather events such as prolonged dry spells and intense rainfall are already affecting rural communities in parts of arid and semi-arid lands (ASALs) of Kenya. At present, nearly 30% of the total human population resides in the

ASALs, which cover approximately 88% of the country's land mass, and hold almost 70% of the total national livestock herd. However, large proportions of pastoralists who reside in ASALs are believed to be at risk of food production deficit, with potential declines in pasture and water availability (Opiyo et al. [2011]), all exacerbated by extreme climate events. Hence, the purpose of this paper is to identify the determinants of vulnerability and measure micro-level vulnerability of pastoralists in arid rangelands of Kenya. The vital information that is obtained from household-level vulnerability analysis is presently lacking in national and regional level assessments. Deressa et al. ([2008]), Pearson et al. ([2008]) and Sherwood ([2013]) show that vulnerability contexts are diverse for different multiple spatial scales, and therefore this approach can contextualize how climate variability and change affect pastoralist livelihoods.

This study was thus carried out to investigate households' vulnerability to climate variability and change to climate-induced stresses in a case study of a pastoralist rangeland of Kenya. The study identifies some of the determining factors for vulnerability based on certain household social, economic and environmental (biophysical) characteristics. The household was selected as the main unit of analysis because major decisions about adaptation to climate-induced stresses and livelihood processes are taken at that level (Thomas [2008]). Nevertheless, households are connected to the wider community, which can greatly influence their decision-making processes in relation to use of particular productive resources. The findings of this study can therefore be useful for targeting interventions, priority setting and resource allocations at micro-level. Complemented with studies analysing climate change impacts and findings from country-level adaptive capacity, governmental policy can be informed. At the same time, the uncertainty associated with climate variability and change demands an approach that prepares rural households without relying on detailed climate projections. The study therefore focused on the adaptive capacity of households for wider resilience programming. Moreover, the findings would assist in resources allocation and determination of resilience investment opportunities that are likely to increase the adaptive capacity of the vulnerable households. The findings in particular suggest potential avenues for research that may further enhance understanding of household vulnerability to climate stresses in the arid rangelands of Kenya.

Study Area

The study was conducted in Turkana County, northwestern rangelands of Kenya (Figure 1). The county lies between longitude 34° 30′ and 36° 40′E and latitude 1° 30′ and 5° 30′ N, covering an area of 77,000 km^2 with an estimated human population of 855,399 according to the 2009 Kenya housing and population census report. The county is one of the poorest in Kenya with an estimated 87.5% of the population living below the absolute poverty line (Kenya National Bureau of Statistics KNBS [2013]). In Lokichoggio, Kakuma and Oropoi divisions where the household survey was undertaken, poverty is exacerbated by the harsh environment, poor infrastructure and low access to basic services, in addition to other underlying causes of poverty that are experienced in northern Kenya. About 96% of the study area falls into arid and very arid eco-climatic zones (Jaetzold and Schmidt [1983]). In the study area, because of aridity, managing short-term climatic fluctuations as well as adapting to long-term changes is critical in sustaining livelihoods. According to the most recent household economy approach assessments, pastoralism is the predominant livelihood accounting for over 55% of the population; 15.6% agro-pastoral, 8% fisher folks and 8.4% are in the urban/peri-urban areas (Household Economy Approach HEA [2012]).

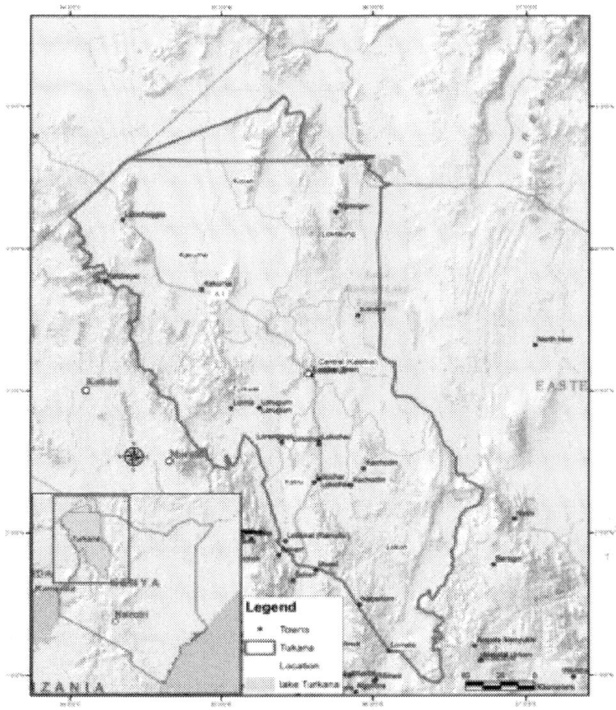

Figure 1: Location of the study area.

Rainfall in the study area exhibits a bimodal pattern, and distribution is unreliable and erratic in both space and time, with an average annual rainfall of 120 mm in the east to over 200 mm in the northwest. Figure 2 shows the annual rainfall variability of the study area. The main rainy season at local level between March 1950 and July 2012 the study area received very low and variable rain. In the past decades, major widespread drought events were experienced in 1980 to 1984, 1990 to 1995, 1999 to 2000, 2008 to 2009 and 2010 to 2011 within the study area. The last drought crisis in the Horn of Africa is estimated to have affected over 13 million people, including 3.75 million Kenyan, especially in the ASALs (UN-OCHA [2011]). The daily temperature ranges are extremely high with an average daily range varying between 24°C and 38°C with a mean of 30°C. The vegetation in this area is mainly scattered *Acacia* bush and a cover of annual herbaceous plants. The density of the woody plants such as *Acacia reficiens* and *A. mellifera* increases on hilly ground.

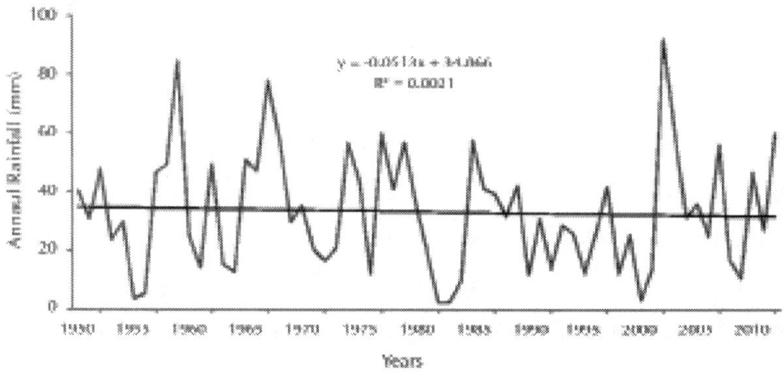

Figure 2: Annual rainfall variability for period 1950 to 2012 in the study area.

The study area is inhabited by the Turkana pastoralists whose major occupation is extensive livestock production (Watson and van Binsbergen [2008]).The livestock species kept include camels, cattle, sheep, goats and donkeys, which have different forage and water requirements and variable levels of resilience during drought episodes. In Turkana, droughts, famine, epidemics and flash floods are also very common occurrences. In most instances, these shocks and stresses are associated with climatic variability and change (Notenbaert et al. [2013]). The outcome of these climatic stresses has been high livestock mortality, loss of property, displacement of people and resources conflicts (Schilling et al. [2012]). In general, by weakening the productivity and functioning of livelihood resources, climatic stresses aggravate the vulnerability of households dependent on these resources. However, Turkana pastoralists are known to be highly mobile with no fixed residence (McCabe et al. [1988]), but their regular patterns of movement often get disrupted by conflicts and disease outbreaks (Hendrickson et al. [1998]; Schilling et al. [2012]).

METHODS

Data Collection

A total of 302 pastoralist households were selected for interviews a using semi-structured questionnaire to elicit responses between June and July 2012. Systematic purposive sampling was used to select drought-prone divisions in Turkana County which include Lokichoggio, Kakuma and Oropoi. However, the specific kraals (homesteads)/ locations where households were selected based on accessibility of the area. Consequently, 10 locations/kraals were selected from which the sample households were selected randomly proportional to population size. Table 1 shows the sampled households from each of the 10 kraals/ locations. A semi-structured questionnaire was used to interview the household heads. The questionnaire was divided into the following: demographic and economic household characteristics, livestock and crops production, access to extension services, credit access, hazards occurrence, perception level, adaptation strategies pursued, different coping strategies, level of resilience and other relevant information.

Table 1: Sampled households by Kraal

Location/ Kraals	Number of households	**Sampled household**	**Percentage of respondent**
Mogila	1,536	20	6.62
Lokichoggio	1,868	9	2.98
Lopusiki	649	1	0.33
Lopwarin	277	11	3.64
Songot	459	19	6.29
Nanam	748	38	12.58
Kakuma	12,767	33	10.93
Letea	4,957	56	18.54
Kalobeyei	2,577	39	12.91
Pelekech	1,863	76	25.17

Kraal used herein to refer to a traditional pastoralists village of huts, typically enclosed by a fence. (Source: Authors' compilation).

Opiyo *et al.*

Opiyo *et al. Pastoralism* 2014 **4**:10, doi:10.1186/s13570-014-0010-9

In addition, *in situ* rainfall and temperature datasets relevant for this study were obtained from Kenya Meteorological Services, National Drought Management Authority (National Drought Management Authority - NDMA [2013]) and County government officers in Lodwar, Turkana to strengthen some of the study findings. The STATA software was used to analyse the data obtained.

Analysis of Households' Vulnerability

There are various ways of analysing vulnerability, namely, socio-economic, biophysical and an integrated approach, which unites both socio-economic and biophysical factors. The socio-economic vulnerability assessment approach focuses on the socio-economic and political status of individuals or groups. Individuals in a community vary in terms of education, gender, wealth, health status, access to credit, access to information and technology, formal and informal (social) capital and political power, which are responsible for variations in vulnerability levels (Füssel[2007]; Deressa et al. [2008]). Consequently, vulnerability is considered to be a starting point or a state that exists within a system before it encounters a hazard event (Kelly and Adger [2000]). In this regard, vulnerability is shaped by society as a result of institutional and economic changes. The socio-economic approach focuses on identifying the adaptive capacity of individuals or communities based on their internal characteristics. One major limitation of this approach is that it focuses only on variations within society, but in reality, societies vary not only due to socio-political factors but also because of environmental or biophysical factors. The socio-economic approach does not account for the natural resource bases which have the potential to counteract the negative impacts of environmental shocks. For example, areas with easily accessible underground water can better cope with drought by utilizing this resource, compared to areas without it (Deressa et al. [2008]).

The second commonly used approach is biophysical that attempts to assess the level of damage that a given environmental stress causes on both social and biological systems. It is sometimes known as an impact assessment. The emphasis is on the vulnerability or degradation of biophysical conditions (Liverman [1990]). This is a dominant approach employed in studies of vulnerability to natural hazards and climate variability and change (Hewitt [1995]). Füssel ([2007]) identified this approach as a risk-hazard approach. The biophysical approach, although very informative, also has a major limitation that assessment of biophysical factors is not a sufficient condition for understanding the complex dynamics of vulnerability. This approach also neglects both structural factors and human agency in producing vulnerability and in coping or adapting to it. The approach overemphasizes extreme events while neglecting root causes and everyday social processes that influence differential vulnerability (Liverman [1990]; Hewitt [1995]; Pulwarty and Riebsame [1997]). The third approach is the integrated vulnerability analysis, which combines both the socio-economic and biophysical factors. This approach includes all the internal state of vulnerability and the external situation. This analytical approach was applied by Madu ([2012]) in agro-ecological based household vulnerability analysis in Ethiopia and by Deressa et al. ([2008]) in regionally based vulnerability analysis.

This present study replicates an integrated vulnerability approach to develop vulnerability indices for each household as proposed by Madu ([2012]) and adopted by Tesso et al. ([2012]) in Ethiopia. In this research, it is assumed that households with higher adaptive capacity are less sensitive to impacts of climate-induced stresses, thus keeping the level of exposure constant. The integrated assessment approach combines both socio-economic and biophysical approaches to determine vulnerability. The vulnerability index development is given as developed by the Intergovernmental Panel on Climate Change IPCC ([2012]) that vulnerability is seen as the net effect of adaptive capacity (socio-economic) and sensitivity/exposure (biophysical):

$$\text{Vulnerability} = (\text{Adaptive capacity}) - (\text{Sensitivity} + \text{Exposure}) \quad (1)$$

When the adaptive capacity of the household exceeds that of its sensitivity and exposure, the household becomes less vulnerable to climate change impacts and the reverse is also true. As explained in the foregoing, each set (adaptive capacity, sensitivity and exposure) is composed of different variables. The model specification further looks like

$$V_i = (A_1X_{1J} + A_2X_{2J} + \cdots + A_{2n}X_{nJ})$$
$$-(A_{n+1}Y_{1J} + A_{n+2}Y_{2J} + \cdots + A_{n+n}X_{nJ}) \quad (2)$$

Where V_i is the vulnerability index, while X_s are elements of adaptive capacity, and Y_s are elements of exposure and sensitivity. The values of X and Y are obtained by normalization using their mean and standard errors. For instance, $X_{IJ} = (X_{IJ} - X_{1*})\ S_1{}^{*}/$, where X_{1j*} is the mean of X_{1j} across the different households, $S_1{}^{*}$ is its standard deviation. X_1 is the principal component result of factors. In this regard, the first principal component of a set of variables is the linear index of all the variables that captures the largest amount of information common to all the variables. The whole matrix of variables of adaptive capacity (X_{ij}) and variables of exposure and sensitivity (Y_{ij}) appears as follows:

$$X_{ij}/_{Y_{ij}} = \begin{pmatrix} (X_{11} + X_{12} + \cdots + X_{2n}) - (Y_{11} + Y_{12} + \cdots Y_{2n}) \\ \cdot \\ (X_{m1} + X_{m2} + \cdots + X_{mn}) - (Y_{m1} + Y_{m2} + \cdots Y_{mn}) \end{pmatrix} \quad (3)$$

The *i* and *j* in the foregoing notation imply the number of rows (in this case is the 302 individual households) and the number of columns (27 variables of adaptive capacity, exposure and sensitivity). In Equation 4, the A_s is the first component score of each variable computed using principal component analysis (PCA) in STATA. Finally, the vulnerability index of each household is obtained using Equation 4 as follows:

$$V_i = \begin{pmatrix} A_1 \\ A_{12} \\ \cdot \\ A_{n+n} \end{pmatrix} \times \begin{pmatrix} (X_{11} + \cdots + X_{2n}) - (Y_{11} + \cdots + Y_{2n}) \\ \cdot \\ (X_{m1} + \cdots + X_{mn}) - (Y_{m1} + \cdots + Y_{mn}) \end{pmatrix} \quad (4)$$

In calculating the direction of relationship in vulnerability indicators (that is, their sign), a negative value was assigned to both exposure and sensitivity. The justification is that households which are highly exposed to climate shocks are more sensitive to damage, assuming constant adaptive capacity. The implication is that a higher net value indicates lesser vulnerability and vice versa. However, in creating the indices, the scale of analysis is important. As noted by Tesso et al. ([2012]) from Deressa et al. ([2008]), vulnerability analysis ranges from local or household level to the global level (Brooks et al. [2005]). The choice of scale is dictated by the objectives, methodologies and data availability.

In this study, the households were classified into three categories based on the value of their vulnerability index, which puts households into highly vulnerable, vulnerable and less vulnerable categories. However, the index computed is not based on the thresholds or presents an absolute value. It is a relative measure, representing the households' own perception of how they have been coping in the past compared to other households.

Factors Influencing Household Vulnerability

Notenbaert et al. ([2013]) note that many factors contribute to vulnerability, and these factors undermine capacity for self-protection, block or diminish access to social protection, delay recovery or expose some households to greater or more frequent hazards than other households. The analysis was performed using ordinal logistic regression analysis. The ordinal logit model is used when the outcome variable is categorized in an ordinal scale, as in this case where vulnerability is ordered as (1) highly vulnerable, which implies households for whom the difference between adaptive capacity and sensitivity/exposure is significantly negative; (2) moderately vulnerable, which means that households for whom the difference between adaptive capacity and sensitivity/exposure is nearly zero; and (3) less vulnerable, which means that the difference between adaptive capacity and exposure/sensitivity is significantly positive. In this study, sensitivity of households to climate-induced stresses is represented by its associated impacts, i.e., shortage of food, loss of water and pasture resources and conflicts faced by those households. In the case of exposure, since all households are assumed to be located in the same environment, exposure is almost uniform across the respondents residing in the study area.

This model is particularly useful in that it can show movement between vulnerability groups, explaining who moves in and out of vulnerability. Following Greene ([1997]), the reduced form of the ordinal logit model is given as

$$Y_j^* = X_j^I \beta + U_{Ij} \tag{5}$$

where Y is the level of vulnerability and involves ordered outcome, that is, $Y=1$ was given to households that have a high level of vulnerability as observed by the negative value of adaptive capacity minus sensitivity/exposure; $Y=2$ was given to households having adaptive capacity nearly equal to their sensitivity/exposure; and $Y=3$ was given to households having their adaptive capacity exceeding their sensitivity and exposure. Y^* is the given state of vulnerability. The X_{ij} are the explanatory variables determining vulnerability level. The independent variables included in the model were sex and age of the household head, experience in the study area, household size and education level of the household head, dependency, marital status, social linkages and visits by extension officers, access to early warning information, non-farm income, herd size, herd structure, access to markets, property regimes, access to remittances, employment, coping strategies, herd diversity, credit access, herd mobility, climate change, experience in increased temperature, drought, floods, and wind, and natural hazards encountered in 5 years. β_S are parameters estimated and U_{ij} is the disturbance term. Y^* is unobserved, but what was observed in this study is

$$Y = 1 \text{ if } Y^* \leq \mu 2 \tag{6}$$

$$Y = 2 \text{ if } \mu 2 < Y^* \leq \mu 3 \tag{7}$$

$$Y = 3 \text{ if } \mu 3 < Y^* \tag{8}$$

Given the cumulative normal function $\Phi\ (\beta' x)$, the probabilities can be shown, thus,

Prob [$y=1$ or highly vulnerable] $= \Phi\ (-\beta' x)$,

Prob [$y=2$ or neutral level of vulnerability] $= \Phi\ (\mu 2 - \beta' x) - \Phi\ (\mu 3 - \beta' x)$

Prob [$y=3$ or less vulnerable] $= 1 - \Phi\ (\mu 3 - \beta' x)$

RESULTS

Hazards Reported by Households

The results show that 49% of households experienced drought over the past three decades, with 27.1% of the respondents reporting livestock disease outbreak, 19% mentioning cross-border inter-community conflicts and 4% citing riverine flash floods as the main hazard (Table 2). However, it was difficult for the respondents to differentiate between threats and hazards, which required clarification of the differences. Drought events were reported to be frequent hazards in the area and had devastating impacts on household livelihoods, pasture and water, which escalates the area's chronic conflicts, insecurity and food insecurity.

Table 2: Hazards identified by respondents

Location/ Kraals	Number of households	Sampled household	Percentage of respondent
Mogila	1,536	20	6.62
Lokichoggio	1,868	9	2.98
Lopusiki	649	1	0.33
Lopwarin	277	11	3.64
Songot	459	19	6.29
Nanam	748	38	12.58
Kakuma	12,767	33	10.93
Letea	4,957	56	18.54
Kalobeyei	2,577	39	12.91
Pelekech	1,863	76	25.17

Kraal used herein to refer to a traditional pastoralists village of huts, typically enclosed by a fence. (Source: Authors' compilation).

Opiyo *et al.*

Opiyo *et al. Pastoralism* 2014 4:10, doi:10.1186/s13570-014-0010-9

Inter-community and ethnic conflict related to scarce grazing land and water resources was reported to have increased inter-tribal animosity, often resulting in armed violent conflicts, which are predatory in nature and much more destructive. Households stated that violent conflicts regularly lead to heavy losses of lives and livelihoods, undermining human and livestock population mobility, as well as development efforts. Study findings show that outbreaks of *peste des petit ruminants* (PPR) locally referred to as *lomoo* is considered as the major hazard, having the highest impact on small ruminants compared to other endemic livestock diseases in the area. On further probing with community animal health workers and veterinary officers, based on the symptoms described by the pastoralist respondents, other diseases were identified as *etome*(mange), *emadang'* (worms infestation), *lukoi* (contagious caprine pleuropneumonia), *lomeri* (lump skin disease), *lokichum* (heart water), *etune* (sheep/goat pox), *lotorebwo or lokipi*(trypanosomiasis), *logooroi* (hemorrhagic septicaemia) and *loukoi* (contagious bovine pleuropneumonia).

Socio-economic and Biophysical Vulnerability

The social and economic variables contributing to vulnerability are summarized in Table 3. Findings show that more than 80% of the respondents had no basic primary education, while the majority are not able to read and write. This in turn reduces a household's ability to understand climatic information, access market and early warning information. Data on household size showed that 64% of those interviewed had more than five persons, with 43% respondents reported more than five dependents. Whereas the sex of the household head is likely to be critical for climate adaptation, the study revealed that more than 50% of households are headed by females. In addition, 54% of the respondents indicated that they do not have access to livestock extension services. The results imply that the vulnerability level of households to the frequently occurring climate-induced stresses is largely determined by gender and education level of the household head in the study area.

Table 3: Social, economic and environmental indicators and their effect on vulnerability level

Hypothesized variables	**Percentage**	**Influence on vulnerabilitya**
Social vulnerability variables		
Sex of HH head: female-headed	50.7	+
Age of HH head: 50+ years	26.6	-
Experiences in the area: less than 5 years	7.9	+
HH size: more than 5 persons	64.6	-
Educational level: no primary education	80.1	+
Dependents: more than 5 persons	43.8	-
Marital status: single (including divorced and widowed)	29.5	+
Linkages: having no social linkages	25.8	-
Visit by extension officers: no access to extension services	54.3	+
Access to early warning information: no access to the information	74.8	+
Economic vulnerability variables		
Non-farm income: have no non-farm income	35.5	-
Herd size: own less than 2 TLUs	18.3	+
Herd structure: no milking herd	72.8	+
Distance to market: more than 10 km away	39.4	-
Property regime: own private land	8.6	-
Access to remittances: no cash transfers	58	+
HH employed: no member of HH employed	88.1	+
HH coping strategies: more than 2 coping strategies	92.1	+
Livestock diversity: less than 2 livestock species	45.4	-
Credit access: having no access to credit at all	77.8	+
Mobility: able to move livestock freely	67.9	-
Environmental vulnerability variables		

Climate change: experiencing change	96.7	+
Temperature: experiencing increase	47.9	+
Drought: noticed increasing events	3.4	+
Flood: noticed change	4.1	+
Wind: noticed unusual change	11.6	+
HH facing more than 2 hazards in 5 years	72.5	–

[a]Positive sign indicates that the variable increases vulnerability, while negative sign means it reduces vulnerability. TLU, tropical livestock unit (1 TLU = 250 kg). (Source: Estimated from June to July 2012 household survey interviews in Turkana; Authors' compilation).

Opiyo *et al.*

Opiyo *et al. Pastoralism* 2014 4:10, doi:10.1186/s13570-014-0010-9

More than 35.5% of the households had no diversified income sources and heavily depend on livestock as their source of income. Results show that only a few of the households were practicing dry land crop farming alongside livestock keeping. For the livestock owned, 18% of respondents had less than two tropical livestock units (TLUs[a]), and 72% of households had no milking herd during the interview period. Further, the result revealed that about 68% of households practice seasonal herd mobility and flexible resource use in the study area. The results show that 88% of households had none of its member formally employed, and 92% of respondents use more than two coping strategies to survive in this harsh climatic condition, despite the low access to credit, access to remittances and long distances to markets. Other economic concern mentioned was the high formal unemployment opportunities, especially for the youths in Turkana.

[a]A TLU is the 250-kg live weight of any domestic herbivore (sheep or goat = 0.1 TLU; donkey = 0.4 TLU; cattle = 0.7 TLU; camel = 1.0 TLU).

Table 3 displays environmental, economic and biophysical factors hypothesized for climate-induced vulnerability. The results show that the sensitivity and exposure to climate variability in terms of those experiencing climate variability (96.7%) and people facing more than two hazards in 5 years (72.5%) have contributed negatively to the vulnerability level of households. Approximately 48% of the respondents noticed an increase in temperature. However, most of the environmental variables increase household's level of vulnerability to climate-induced stresses as indicated by the positive sign.

Measuring Household-level Vulnerability

Table 4 presents the result of principal component analysis and its association with the social, economic and environmental variables. The factor scores (weights) of the first principle component analysis was positively associated with the majority of the indicators identified under adaptive capacity, exposure and sensitivity. Holding exposure and sensitivity constant, a negative index shows the household to have relatively lower adaptive capacity when compared to a household with a positive index value and vice versa.

Table 4: Factor score for the first principal component analysis

Factors	**Factor Score**
Social vulnerability variables	
Sex of HH head: female-headed	0.22917
Age of HH head: 50+ years	−0.25804
Experiences in the area: less than 5 years	−0.02906
HH size: more than 5 persons	−0.29837
Educational level: no primary education	0.039479
Dependents: more than 5 persons	−0.31599
Marital status: single (including divorced and widowed)	0.07224
Linkages: having no social linkages	−0.04414
Visit by extension officers: no access to extension services	0.189862
Access to early warning information: having no access	0.19573
Economic variables	
Non-farm income: HH with no farm income	−0.34130
Herd size in TLU: own less than 2 TLUs	0.293813

Herd structure: no milking herd	0.100562
Distance to markets: more than 10 km away	−0.15870
Property regime: own private land	−0.06728
Access to remittances: no cash transfers	0.047677
HH employed: no member of HH employed	0.095173
HH coping strategies: more than 2 coping strategies	0.101335
Livestock diversity: less than 2 livestock species	−0.42800
Credit access: having no access to credit at all	0.178153
Mobility: able to move livestock freely	−0.14718
Environmental variables	
Climate change: experiencing change	−0.02276
Temperature: experiencing increase	0.081161
Drought: noticed increasing events	0.164723
Flood: noticed change	0.039066
Wind: noticed unusual change	0.259975
HH facing more than 2 hazards in 5 years	−0.06667

Abbreviations: Households (HH), Tropical livestock unit (TLU).

Opiyo *et al.*

Opiyo *et al. Pastoralism* 2014 4:10, doi:10.1186/s13570-014-0010-9

To compute the vulnerability index in Equation 2, indicators of adaptive capacity, which are positively associated with the first principal component analysis, and indicators of sensitivity and exposure, which are negatively associated with the principal component analysis, were used in this study. The variables considered in Equation 2 include sex of the household head, education level, marital status, access to extension services and early warning information, livestock ownership, herd structure, access to cash remittances, household employment status, coping strategies and access to credit. However, for the exposure and sensitivity, all the variables were considered in the analysis. This is because adaptive capacity is considered as positively contributing to the reduction of vulnerability, while exposure and sensitivity are negatively contributing to vulnerability reduction. The larger the factor score the more important is the variable and contributes more to the household's vulnerability.

The households were classified into three categories using the vulnerability index: less vulnerable are households that are in a vulnerable situation but can still cope; moderately vulnerable households are those that need urgent but temporary assistance in case of shock and stresses; and the highly vulnerable are those households that are almost at a point of no return. The result shows that the majority of households fall within the moderately vulnerable category, with 44% households having an index from −1.00 to 1.00. The less vulnerable households had an index of 1.1 to 3.0 and constitute 29%, while the highly vulnerable households had an index of −0.9 to −3.0 but are 27% of the total households sampled (Table 5 and Figure 3). Although there seems to be normal vulnerability index distribution in Figure 3, a keen look at the values showed a slight shift to the left. Statistically, the households' distribution is skewed toward the highly vulnerable in the illustration. In general, the results reveal high and moderate vulnerability levels of the pastoralist households in Turkana.

Table 5: Classification of community according to the range of their vulnerability index

Vulnerability category	Household situation	Vulnerability index	Percentage of households
Highly vulnerable	Emergency level HHs	−0.9 to −3.5	27
Moderately vulnerable	Needs urgent but temporary external assistance to recover	−1.0 to +1.0	44
Less vulnerable	In a vulnerable situation but still able to cope	+1.1 to +3.0	29
Total			100

(Source: Computed from household questionnaire interviews in June to July 2012).

Opiyo *et al.*

Opiyo *et al.* *Pastoralism* 2014 4:10, doi:10.1186/s13570-014-0010-9

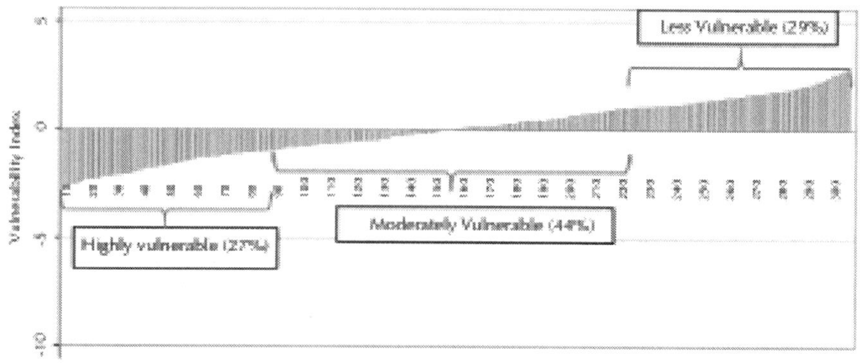

Figure 3: Vulnerability index by households in Kakuma, Oropoi and Lokichoggio divisions in Turkana.

Determinants of Vulnerability

The majority of highly vulnerable households were headed by females or someone with no primary level of education (Table 6). The highly vulnerable households also had more than five dependents, no access to early warning information, no milking herd and own less than two livestock species, and perceived changes in climate. By contrast, a household is likely to be less vulnerable when they are headed by a male, the household head is literate, not divorced or widowed, when they have access to extension services and early warning information, and own large and mixed-species herds. Similarly, households with access to cash remittances, with a member of the household in formal employment, and those with diverse coping strategies and access to credit facilities are reported to be less vulnerable to climate extremes. The results suggest that because of scarce resources during extreme climate events, households with high number of dependents and without any other livelihood diversification activities are likely to be more vulnerable than households with complementary sources of income and less dependents.

Table 6: Statistical description of model variables by vulnerability category

Vulnerability variables	Vulnerability category		
	Less vulnerable	Moderate vulnerable	Highly vulnerable
Sex of HH head: female	67 (44%)	62 (20%)	24 (45%)
Age of HH head: 50+ years	6 (41%)	33 (34%)	35 (25%)
Experiences in the area: less than 5 years	10 (16%)	10 (44%)	4 (4%)
HH size: more than 5 persons	27 (8%)	92 (22%)	76 (42%)
Educational level: no primary education	72 (45%)	107 (5%)	63 (55%)
Dependents: more than 5 persons	10 (47%)	58 (41%)	59 (33%)
Marital status: single (including divorced and widowed)	31 (42%)	41 (55%)	17 (46%)
Linkages: HH with no social linkages	18 (42%)	34 (52%)	26 (21%)
Visit by extension officers: no access to extension services	67 (17%)	65 (39%)	32 (20%)
Access to early warning information: having no access	77 (14%)	100 (9%)	49 (45%)
Non-farm income: HH with no farm income	5 (47%)	43 (33%)	58 (34%)
Herd size in TLU: own less than 2 TLUs	23 (39%)	17 (44%)	4 (31%)
Herd structure: no milking herd	72 (30%)	97 (23%)	51 (42%)
Distance to market: more than 10 km away	20 (44%)	51 (18%)	40 (27%)
Property regime: own private land	4 (26%)	13 (46%)	9 (40%)
Access to remittances: no cash transfers	58 (8%)	71 (36%)	46 (30%)
HH employed: no member of HH employed	81 (46%)	118 (15%)	67 (30%)

HH coping strategies: more than 2 coping strategies	82 (46%)	121 (50%)	66 (25%)
Livestock diversity: less than 2 species	5 (35%)	57 (35%)	75 (58%)
Credit access: having no access to credit at all	78 (46%)	108 (33%)	49 (17%)
Mobility: able to move livestock freely	42 (19%)	93 (41%)	70 (32%)
Climate change: experiencing change	43 (23%)	59 (26%)	38 (47%)
Drought: noticed increasing events	4 (44%)	3 (30%)	3 (21%)
Flood: noticed change	3 (33%)	7 (44%)	2 (29%)
Wind: noticed unusual change	11 (41%)	16 (25%)	7 (45%)
HH facing more than 2 hazards in 5 years	63 (40%)	99 (30%)	57 (26%)

HH, households; TLU, tropical livestock unit.

Opiyo *et al.*

Opiyo *et al. Pastoralism* 2014 4:10, doi:10.1186/s13570-014-0010-9

As shown in Table 6, the majority of households in the moderately vulnerable category are the ones with less than 5 years' experience, either divorced or widowed household heads, household heads with no social linkages, household heads with no access to extension services, households who own less than two TLUs, own private lands, households which do not received any cash remittances, household heads with more than two coping strategies, households who practice mobility and are able to move freely with their livestock herd, and perceive climatic changes in the area.

The results of the ordered logistic regression model for all the single predictors' variables influencing a household's vulnerability are summarized in Table 7. A number of variables were statistically significant at 1% and 5% levels of significance and were important in influencing households' vulnerability to climate-induced stresses. The variables that showed significant influence on vulnerability include gender of the household head, age of the household head, size of the household, number of dependents, marital status, access to extension services and early warning information, complementary sources of income, herd diversity and structure, coping strategies and access to credit.

Table 7: Factors influencing household's vulnerability

Variables	Estimate	SE	OR	*z*	*pvalue*
Sex of HH head: female	−1.3828	0.2303	0.2509	−6.0049	<0.0001*
Age of HH head: 50+ years	1.5714	0.2704	4.8134	5.8106	<0.0001*
Experiences in the area: less than 5 years	−0.6187	0.399	0.5386	−1.5506	0.1220
HH size: more than 5 persons	2.2155	0.2651	9.1663	8.3585	<0.0001*
Educational level: no primary education	−0.1754	0.2684	0.8391	−0.6534	0.5140
Dependents: more than 5 persons	2.1434	0.2639	8.5281	8.1212	<0.0001*
Marital status: single (including divorced and widowed)	−0.4689	0.2349	0.6257	−1.9965	0.0468*
Linkages: having no social linkages	0.4232	0.2456	1.5268	1.7233	0.0859**
Visit by extension officers: no access to extension services	−1.0717	0.2237	0.3424	−4.7909	<0.0001*
Access to early warning information: having no access	−1.017	0.2535	0.3617	−4.0119	<0.0001*
Non-farm income: HH with no farm income	2.3332	0.2715	10.3108	8.5946	<0.0001*
Herd size in TLU: own less than 2 TLUs	−1.8785	0.3414	0.1528	−5.502	<0.0001*
Herd structure: no milking herd	−0.668	0.2434	0.5128	−2.7442	0.0064*

Distance to market: more than 10 km away	0.7694	0.2301	2.1586	3.3442	<0.0001*
Property regime: own private land	0.5666	0.375	1.7623	1.5111	0.1318
Access to remittances: no cash transfers	−0.2696	0.2161	0.7637	−1.2477	0.2131
HH employed: no member of HH employed	−0.6226	0.3337	0.5365	−1.866	0.0630**
HH coping strategies: less than 2 coping strategies	−1.0896	0.4268	0.3363	−2.5529	0.0112*
Livestock diversity: less than 2 livestock species	3.3875	0.3506	29.592	9.663	<0.0001*
Credit access: having no access to credit at all	−1.1825	0.2703	0.3065	−4.3753	<0.0001*
Mobility: able to move livestock freely	1.2984	0.2424	3.6635	5.3571	<0.0001*
Temperature: experiencing increase	−0.0412	0.2175	0.9597	−0.1892	0.8500
Drought: noticed increasing events	−0.2123	0.6269	0.8087	−0.3387	0.7350
Flood: noticed change	−0.1252	0.5225	0.8823	−0.2396	0.8108
Wind: noticed unusual change	−0.2462	0.3355	0.7818	−0.7338	0.4636
HH facing more than 2 hazards in 5 years	−0.0369	0.2409	0.9638	−0.1532	0.8784

SE, standard error; OR, odds ratio; *z*, score of two sample tests (basically a *t* test) though using the standard normal to calculate the *p*value and expressing the statistically significance levels at *5% and **10%. (Source: Ordinal logistic regression model: single predictors).

Opiyo *et al.*

Opiyo *et al. Pastoralism* 2014 4:10, doi:10.1186/s13570-014-0010-9

The higher odds ratios computed for age of the household head, household size, dependents, complementary income sources, herd diversity and mobility imply that these factors have more effects on the households' vulnerability level. But it should be noted that majority of the sampled population derived their means of subsistence from nomadic pastoral production system which are weather sensitive and therefore absence of non-agricultural income sources (salary or wages labour) which are not weather sensitive would be a significant determinant on vulnerability level as reflected in the odds ratio figures. Herd diversity and mobility allow households to spread risk and pursue various coping strategies. Herd mobility was observed to be a significant (at 5%) determinant of households' vulnerability to climate variability and change.

From the ordered regression results, 75% of female-headed households are more likely to be vulnerable to climate-induced stresses and shocks compared to the male-headed households. This might be because female-headed households normally face gender discrimination with respect to resources, rights, education, income and economic opportunities. Household age was also an important demographic factor determining how vulnerable a household could be. For example, households headed by person above 50 years of age are more likely to be vulnerable compared with the younger persons. Consequently, elderly household heads are probably worse off in terms of preparing strategies to cushion their families against adverse climatic stresses and impacts and likely to make them more vulnerable. Similarly, the more dependents a household has the more likely for it to be vulnerable since a larger proportion of household resources are directed to dependents who cannot contribute much toward household welfare. Further results also show that widowed or divorced single-headed families are 37.4% more likely to be vulnerable than families where both spouses are present. According to the study results, it was observed that livestock diversity had higher odds ratio of 29.592, which suggest that households with more than two livestock species are 28 times more likely to shift from highly vulnerability category to moderately vulnerability level.

Although 27 variables were hypothesized to be correlated with vulnerability, ordinal regression model result confirmed that only 17 factors were significant (at $p<0.1$ and $p<0.05$) in influencing households' vulnerability. Although not computed in this present study, it should be noted that different determinants have differential effects on a household's vulnerability levels.

DISCUSSION

Drought was considered by the respondents as one of the most frequent hazards in Turkana, in addition to other hazards such as conflicts, disease outbreaks and flooding. The pastoralist respondents maintained that frequent, prolonged drought events have increased in severity over the past decade or so. This perception corroborate Nicholson's ([2014]) assertion about drought events in northern Kenya, with rainfall being at least 50% to 75% below normal in almost half of the drought-stricken region and is likely to be linked to changing climatic conditions within the greater Horn of Africa. However, there is a great deal of uncertainty and tension about when, where and how much climate change will happen. Even less is known about the actual impacts of these perceived changes (Kabubo-Mariara [2009]; Notenbaert et al. [2013]; Intergovernmental Panel on Climate Change IPCC [2014]) to the dry land communities. It is argued that due to the frequency of shocks in the study area, adaptive or coping capacity and mechanisms adopted by vulnerable households could well have equal or larger ranges to that of exposure and sensitivity. As discussed earlier on vulnerability, the interaction of environmental (biophysical) forces determines exposures and sensitivities, and various social, cultural, political and economic forces shape adaptive capacity (Yohe and Tol [2002]; Turner et al. [2003]; Skjeflo [2013]).

In the study area just like the rest of northern Kenya, pastoralist households largely bear the brunt of negative impacts of extreme climate events like drought, which include increased poverty, water scarcity, resource-based conflicts, disease outbreaks and food insecurity. The majority of people in Turkana live below the absolute poverty line (an estimated 87.5% of the population) with more than 50% heavily relying on food aid and safety net programmes from year to year. The people who are already poor in this remote part of the country are struggling to cope with the added burden of increasingly unpredictable weather, which is triggered by climate variability and change. Evidence shows that Turkana is constrained by the harsh climatic conditions and remoteness, coupled with poor infrastructure and low access to essential services (Republic of Kenya [2012]) Previous studies (Thornton et al. [2006]; Kabubo-Mariara [2009]) in the region confirmed that it is becoming increasingly difficult for households to bounce back from ever-changing, inconsistent weather affecting their livelihoods, and many have been forced to pursue other livelihoods and coping mechanisms that only increase the cycle of vulnerability.

This study has shown that female-headed households, households with experience of less than 5 years in the area, household heads with no primary level of education and households headed by divorced and widowed persons, with no access to extension services and early warning information, in particular, are disproportionately likely to be affected by climate stresses and variability. In times of climate stresses and shocks like drought, these categories of households tend to have fewer options to find other ways of making a living, because their very low levels of literacy reduce their opportunities in coping mechanisms such as wage employment. Similarly, female or divorced and widowed household heads are likely not to be empowered enough in pastoral communities to make household decisions (Nabikolo et al. [2012]) and are frequently without access to credit services and adequate capital assets or not able to own large herds to manage households' daily requirements. Similar observations have been made by Kakota et al. ([2011]) in Malawi and Tesso et al. ([2012]) in Ethiopia that widowed or divorced household heads are more vulnerable because they rely on income earned by either the father or mother as the bread winners. These findings make a strong case for continuous targeting of pastoralist women in resilience-building interventions in the rangelands.

The results are consistent with previous findings (Deressa et al. [2008]) in similar ecosystems. However, for the biophysical variables, the greater the level of household reliance on natural resources, such as pastoralism or dry land crop farming, the greater will be their vulnerability to climate variability and change. This is partly because the use of such natural resources is dependent on rainfall, which is projected to change. This study observed that almost all the postulated biophysical/environmental variables contribute positively to household vulnerability. It is likely that the level of dependence on natural resources especially pastures and water will vary from household to household. For example, while the majority of households (78.9%) depend on livestock herding as their main source of livelihood, for others, livestock is just an equal or lesser contributor besides other economic activities.

The level of vulnerability is closely associated with the degree of poverty in Turkana. This county has higher degrees of vulnerability in terms of households' adaptive capacity or social and economic characteristics that are linked to their exposure and sensitivity to biophysical variables (Lo´pez-Carr et al. [2014]). The findings of this study suggest that in case of an external stress or shock, the majority of the households would need some assistance for them to recover. These results reflect the findings of studies by Kenya National Bureau of Statistics (KNBS) ([2013]) which showed that Turkana is the poorest county in Kenya, with 87.5% of the population living in absolute poverty. Other studies have indicated that Turkana region has a number of households with high of dependents, low engagement in off-farm activities and low levels of education (Blench[2000]; McPeak and Barrett [2001]; Watson and van Binsbergen [2008]).

The determinants of households' vulnerability were found to be significantly influenced by the sex of the household head, age of the household head, size of the household, number of dependents, marital status, social linkages, access to extension services and early warning information. In addition, non-farm income, herd size and diversity, herd structure and herd mobility, access to markets, households' employment status, coping strategies and access to credit were also observed to be the key determinants of the households' vulnerability to climate-induced stresses. This concurs with studies by Eriksen et al. ([2005]) and Notenbaert et al. ([2013]) which similarly observed some of these factors to be the key determinant of households' vulnerability to climate variability and change in rural communities. The results are also consistent with previous findings by Kakota et al. ([2011]) and Gebrehiwot and van der Veen ([2013]). From these findings, it seems there is still more to be done to understand vulnerability and its underlying processes. In this paper, the causal relationships between the statistically significant variables and outcomes in terms of vulnerability levels are explained using existing literature. Some of these explanations are, however, rather speculative and not confirmed for the local situation. More in-depth qualitative fieldwork, such as including open-ended questions, conducting focus-group discussions or in-depth interviews with selected households based on these factors, could strengthen our confidence in some of these explanations. The combination of quantitative surveys with qualitative autopsy is likely to provide a much more in-depth analysis of determinants of vulnerability.

CONCLUSIONS

Understanding vulnerability to environmental change and extreme climate events is necessary for policy makers to develop mitigation and adaptation programmes for long-term resilience. Vulnerability analyses contribute to the knowledge on climate-sensitive socio-economic or ecological systems, enabling policy to be targeted on the most vulnerable places, sectors or people and adaptation options to be defined. The results of this study from Turkana portray social and biophysical vulnerability indices which are useful to local development programming for long-term resilience. However, as a prerequisite for building households' resilience to climatic extremes, in-depth understanding is necessary of the adaptive capacity, exposure and sensitivity.

This study focused on the micro-level to assess the variability of vulnerability across different households. Categorization of vulnerability levels was used to help identify households that are not vulnerable currently but have a high probability of becoming vulnerable in the future. In the future, studies assessing household-level vulnerability to climate-induced stresses should explore the use of panel data as well as cross-sectional data to portray longitudinal and cross-sectional characteristics of households. Similarly, future resilience interventions should target individual households within a community because major decisions about adaptation to climate-induced stresses and livelihood processes are taken at the micro-level.

We conclude that because of various social, economic and biophysical determinants observed to influence households' vulnerability to climate-induced stresses, policies with emphasis on women's empowerment, promoting education, supporting extension services and enhancing diversifications of income sources and access to credit, supporting herd mobility and diversity, creating employments, and increasing access to markets and early warning information are likely to improve resilience of pastoral households. Although the results of this study are specific to Turkana County in Kenya, the approach and findings could be applicable to other arid and semi-arid areas in the region.

AUTHORS' CONTRIBUTIONS

FEOO collected the data and drafted the manuscript. OVW and MN guided the design of the research study and provided technical reviews of the paper. All authors read and approved the final manuscript.

ACKNOWLEDGEMENTS

The work has been supported by small grants from the African Climate Change Fellowship Program, Global Change for System Analysis, Research & Training (ACCFP/START), and additional support from the Red Cross/Red Crescent Climate Centre/START and United Nations International Strategy for Disaster Reduction (UNISDR) under Climate and Development Knowledge Network small grant. The authors also thank Boniface Makone of the Welcome Trust for the statistical guidance during the preparation of this paper.

REFERENCES

1. Adger, WN, and PM Kelly. 1999. Social vulnerability to climate change and the architecture of entitlements. Mitigation and Adaptation Strategies for Global Change 4: 253–266.
2. Adger, WN. 1999. Social vulnerability to climate change and extremes in coastal Vietnam. World Development 27: 249–269.
3. Adger, WN. 2006. Vulnerability. Global Environmental Change 16 (3): 68–281.
4. Blench, R. 2000. Extensive pastoral livestock systems: Issues and options for the future. GCP/JPN/005/ JPN. FAO (Food and Agriculture Organization of the United Nations), Rome, Italy.
5. Brooks, N, WN Adger, and PM Kelly. 2005. The determinants of vulnerability and adaptive capacity at the national level and the implications for adaptation. Global Environmental Change 15 (2): 151–163.
6. Callaway, JM. 2004. Adaptation benefits and costs:how important are they in the global policy picture and how can we estimate them? Global Environmental Change 14: 273–284.

7. Carpenter, S, B Walker, JM Anderies, and N Abel. 2001. From metaphor to measurement: resilience of what to what? Ecosystems 4 (8): 765–781.
8. Deressa, T, RM Hassan, and C Ringler. 2008. Measuring Ethiopian farmers' vulnerability to climate change across regional states, IFPRI discussion paper no. 806.
9. Eriksen, S, and K O'Brien. 2007. Vulnerability, poverty and the need for sustainable adaptation measures. Climate Policy 7: 337–352.
10. Eriksen, S, K Brown, and K Mick. 2005. The dynamics of vulnerability:locating coping strategies in Kenya and Tanzania. Geographical Journal 171: 287–305.
11. Fraser, EDG, AJ Dougill, K Hubacek, CH Quinn, J Sendzimir, and M Termansen. 2011. Assessing vulnerability to climate change in dryland livelihood systems:Conceptual challenges and interdisciplinary solutions. Ecology and Society 16 (3): 14.
12. Füssel, H. 2007. Vulnerability:a generally applicable conceptual framework for CC research. Global Environmental Change 17: 155–167.
13. Gallopın, GC. 2006. Linkages between vulnerability, resilience, and adaptive capacity. Global Environmental Change 16: 293–303.
14. Gebrehiwot, T, and A van der Veen. 2013. Climate change vulnerability in Ethiopia:disaggregation of Tigray Region. Journal of Eastern African Studies 7 (4): 607–629.
15. Greene, WH. 1997. Econometric analysis. Prentice Hall, New Jersey.
16. Hendrickson, D, J Armon, and R Mearns. 1998. The changing nature of conflict and famine vulnerability: the case of livestock raiding in Turkana District, Kenya. Disasters 22: 185–199.
17. Herrero, M, C Ringler, J van de Steeg, P Thornton, T Zhu, E Bryan, A Omolo, J Koo, and A Notenbaert. 2010. Climate variability and climate change and their impacts on Kenya's agriculture sector. International Livestock Research Institute (ILRI), Nairobi, Kenya.
18. Hewitt, K. 1995. Excluded perspectives in the social construction of disaster. International Journal of Mass Emergencies and Disasters 13: 317–319.

19. Hinkel, J. 2011. Indicators of vulnerability and adaptive capacity: towards a clarification of the science–policy interface. Global Environmental Change 21: 198–208.
20. 2012. Livelihood profiles:Six livelihood zones in Turkana County, Kenya. June 2012 report prepared for Oxfam and Save the Children UK.
21. 2001. Climate change 2001:Impacts, adaptations and vulnerability. Report of working group II. Synthesis report. Cambridge University Press, Cambridge.
22. 2012. Managing the risks of extreme events and disasters to advance climate change adaptation. A special report of Working Groups I and II of the Intergovernmental Panel on Climate Change. Cambridge University Press, Cambridge, UK.
23. 2014. Climate change: Impacts, adaptation, and vulnerability. Working Group II contribution to the IPCC 5th assessment report.
24. Jaetzold, R, and H Schmidt. 1983. Farm management in Kenya, vol. II part C: East Kenya. Ministry of Agriculture, Nairobi, Kenya.
25. Kabubo-Mariara, J. 2009. Global warming and livestock husbandry in Kenya:impacts and adaptations. Ecological Economics 68: 1915–1924.
26. Kakota, T, D Nyariki, D Mkwambisi, and W Kogi-Makau. 2011. Gender vulnerability to climate variability and household food insecurity. Climate and Development 3 (4): 298–309.
27. Kasperson, J, R Kasperson, and B Turner. 1995. Regions at Risk: Comparisons of Threatened Environments. United Nations University Press, New York.
28. Kelly, PM, and WN Adger. 2000. Theory and practice in assessing vulnerability to climate change and facilitation adaptation. Climatic Change 47 (4): 925–1352.
29. 2013. Exploring Kenya's inequality: Pulling apart or pooling together. National Bureau of Statistics and Society for International Development (SID), Nairobi, Kenya.
30. Klein, RJ, S Eriksen, LO Naess, A Hammill, TM Tanner, C Robledo, and K O'Brien. 2007. Portfolio screening to support the mainstreaming of adaptation to climate change into development assistance, Tyndall Centre Working Paper No. 102.

31. Liverman, DM, et al.. 1990. Vulnerability to global environmental change. In Understanding global environmental change, ed. Kasperson RE, 27–44. Center for Technology, Environment, and Development, Clark University, Worcester.

32. Lo´pez-Carr, D, GP Narcisa, EA Juliann, MM Jankowska, C Funk, G Husak, and J Michaelsen. 2014. A spatial analysis of population dynamics and climate change in Africa: potential vulnerability hot spots emerge where precipitation declines and demographic pressures coincide. Population Environment 35: 323–339.

33. Madu, IA. 2012. Partial vulnerability of rural households to climate change in Nigeria:Implication for internal security. Working paper number 2; May 2012. The Robert S. Strauss Center for International Security and Law, University of Texas, USA.

34. Marshall, NA, CJ Stokes, NP Webb, PA Marshall, and AJ Lankester. 2014. Social vulnerability to climate change in primary producers: a typology approach. Agriculture, Ecosystems and Environment 186: 86–93.

35. McCabe, JT, R Dyson-Hudson, PW Leslie, PH Fry, N Dyson-Hudson, and J Wienpahl. 1988. Movement and migration as pastoral responses to limited and unpredictable resources. In Arid lands today and tomorrow, ed. Whitehead EE, Hutchinson CF, Timmerman BN, Varady RG, 727–734. Proceedings of an International Research and Development Conference, Tucson, Arizona, USA.

36. 2001. Climate change 2001: Impacts, adaptation and vulnerability. Cambridge University Press, Cambridge.

37. McPeak, JG, and CB Barrett. 2001. Differential risk exposure and stochastic poverty traps among east African pastoralists. American Journal of Agricultural Economics 83 (3): 674–679.

38. Nabikolo, D, B Bashaasha, MN Mangheni, and JGM Majaliwa. 2012. Determinants of climate change adaptation among male and female headed farm households in eastern Uganda. African Crop Science Journal 20 (2): 203–212.

39. 2013. Turkana County drought monitoring and early warning bulletin - December 2013.

40. Nicholson, SE. 2014. A detailed look at the recent drought situation in the Greater Horn of Africa. Journal of Arid Environments 104: 71–79.

41. Notenbaert, A, SN Karanja, M Herrero, M Felisberto, and S Moyo. 2013. Derivation of a household-level vulnerability index for empirically testing measures of adaptive capacity and vulnerability. Regional Environmental Change 13: 459–470.
42. Opiyo, FEO, SM Mureithi, and RK Ngugi. 2011. The influence of water availability on pastoralist's resource use in Kitui and Mwingi districts of Kenya. Journal of Human Ecology 35 (1): 43–52.
43. Paavola, J. 2008. Livelihoods, vulnerability and adaptation to climate change in Morogoro, Tanzania. Environmental Science and Policy 11 (7): 642–654.
44. Pearson, L, R Nelson, S Crimp, and J Langridge. 2008. Climate change vulnerability assessment: Review of agricultural productivity. CSIRO climate adaptation flagship working paper 08.1.
45. Pulwarty, RS, and WE Riebsame. 1997. The political ecology of vulnerability to hurricane-related hazards. In Hurricanes: Climate and socioeconomic impacts, 185–214. Springer, Berlin, Heidelberg, New York.
46. 2012. National policy for sustainable development of Northern Kenya and other arid lands; Sessional Paper No. 8 of 2012, Republic of Kenya.
47. Schilling, J, FEO Opiyo, and J Scheffran. 2012. Raiding pastoral livelihoods: Motives and effects of violent conflict in north-western Kenya. Pastoralism: Research, Policy and Practice 2: 25.
48. Sherwood, A. 2013. Community adaptation to climate change: exploring drought and poverty traps in Gituamba location, Kenya. Journal of Natural Resources Policy Research 5 (2–3): 147–161.
49. Silvestri, S, E Bryan, C Ringler, M Herrero, and B Okoba. 2012. Climate change perception and adaptation of agro-pastoral communities in Kenya. Regional Environmental Change 12 (4): 791–802.
50. Skjeflo, S. 2013. Measuring household vulnerability to climate change—Why markets matter? Global Environmental Change 23 (6): 1694–1701.
51. Stringer, LC, MS Reed, AJ Dougill, and C Twyman. 2009. Local adaptations to climate change, drought and desertification: Insights to enhance policy in southern Africa. Environmental Science and Policy 12: 748–765.

52. Tesso, G, B Emana, and M Ketema. 2012. Analysis of vulnerability and resilience to climate change induced shocks in North Shewa, Ethiopia. Agricultural Sciences 3: 871–888.

53. Thomas, RJ. 2008. Opportunities to reduce the vulnerability of dryland farmers in Central and West Asia and North Africa to climate change. Agriculture, Ecosystems and Environment 126 (1–2): 36–45.

54. Thornton, PK, PG Jones, T Owiyo, RL Kruska, M Herrero, P Kristjanson, A Notenbaert, N Bekele, and A Omolo. 2006. Mapping Climate Vulnerability and Poverty in Africa. Report to the Department for International Development, International livestock research institute, Nairobi, Kenya.

55. Turner, BL, RE Kasperson, PA Matson, JJ McCarthy, RW Corell, L Christensen, N Eckley, JX Kasperson, A Luers, ML Martello, C Polsky, A Pulsipher, and A Schiller. 2003. A framework for vulnerability analysis in sustainability science. Proceedings of the National Academy of Sciences of the United States of America 100 (14): 8074–8079.

56. 2009. Global assessment report on disaster risk reduction.

57. 2004. Living with risk: A global review of disaster reduction initiatives. United Nations International Strategy for Disaster Reduction, Geneva, Switzerland.

58. 2011. Horn of Africa Crisis: Situation Report No. 13: United Nations Office for Co-ordination of Humanitarian Affairs.

59. Watson, DJ, and J van Binsbergen. 2008. Livestock market access and opportunities in Turkana, Kenya. ILRI Research Report 3, 106. ILRI (International Livestock Research Institute), Nairobi, Kenya.

60. Yohe, G, and RSJ Tol. 2002. Indicators for social and economic coping capacity—moving toward a working definition of adaptive capacity. Global Environmental Change 12: 25–40.

61. Yuga, NG, PG Shivakoti, and RP Sylvain. 2010. Household-level vulnerability to drought in hill agriculture of Nepal: Implications for adaptation planning. International Journal of Sustainable Development and World Ecology 17 (3): 225–230.

Chapter 7

Global Health and Natural Disaster Alerts: Preparing Mobile Phones to Endure the Unthinkable

Wladimir J Alonso[1, 2], Cynthia Schuck-Paim[2], and Ghassem R Asrar[3]

[1]Fogarty International Center, National Institutes of Health (NIH), 16 Center Drive, Building 16, 20892 Bethesda, MD, USA

[2]Origem Scientifica, Sao Paulo, Brazil

[3]Joint Global Change Research Institute, Pacific Northwest National Laboratory, University of Maryland, Maryland, MD, USA

ABSTRACT

Despite the increasingly positive role of portable communication technologies for socioeconomic development and their growing use in global health and other emergency contexts, several challenges still hinder exploring the full potential of mobile phones as effective mitigation tools in natural disasters, public health emergencies and in the aftermath of extreme disruptive events. Mobile devices are designed and advertised to withstand predominantly the demands of normal daily situations, being fraught with fragilities that limit their utility for effective communication and coordination of help in emergency situations. We discuss ways to overcome some of these limitations in the future by the incorporation of features to increase their resilience and effectiveness as aid tools at relatively low cost. Improvements in autonomous energy generation and use, based on existing and rapidly emerging technologies, as well as further improvements in physical durability and off-line operability are encouraged. We also identify the possibility to combine capabilities from other devices, such as space-based telecommunication systems and traditional two-way radios, to enhance the utility of mobile devices for these applications. The solutions we propose can help millions of citizens around the world to manage the risks and impacts of natural and health-related hazards. They should also promote further resilience to avoiding and recovering from such events, especially in vulnerable regions with limited infrastructure.

REPORT

We just celebrated the 40th anniversary of the first call made from a mobile phone (BBC 2013). Since then, an extraordinary technological revolution has been unleashed, enabling three-quarters of the world population (and much of the developing world) to have access to this powerful communication device at all times (The World Bank 2012). Nowadays, more people in the world have access to mobiles than to basic sanitation. Almost everywhere, individuals are in constant contact with peers, family, business partners, service providers and an ever growing number of information sources. The rippling effects brought about by these ubiquitous communication opportunities on the living standards of populations include the creation of job opportunities, the promotion of political transparency, the early containment of epidemic outbreaks, the early warning of hazards and the possibility to scale up global health systems (The World Bank 2012; The Economist 2009a; The Economist 2009b; Quadir 2005; Lester & Karanja 2008; BBC Media Action 2012; The Economist 2007).

Mobile phones have also played an increasingly positive role as disaster relief tools. Since the year 2000, natural disasters and extreme weather events have caused over one million deaths and directly affected more than 2 billion people (Guha-Sapir et al. 2013). In addition to mass injury and death, common features of such disasters include damage to infrastructure, difficult access to basic goods, overburden of emergency and medical services and frequent isolation of victims. Although we currently count on sophisticated forecast and response systems (such as early assessment of hurricane paths or of the likelihood of a new disease spreading), new vulnerabilities are also present due to the increasing complexity of infrastructures and rapid expansion of megacities (especially in regions that are geologically and/or meteorologically vulnerable), interdependency of technologies and their massive use by growing populations (Kenett & Portugali2012). In such contexts, the use of mobile phones can be critical, particularly in light of the likely collapse of terrestrial telecommunication infrastructures following a disaster (Patricelli et al. 2009; Townsend & Moss 2005).

However, several challenges still constrain the greater potential that mobile phones offer as effective mitigation tools in the aftermath of disasters and in other emergencies. For example, during earthquakes, hurricanes, floods, pandemics and other severe disease outbreaks, economic, societal and personal infrastructures are severely affected and governmental capabilities become quickly saturated due to the sheer demand for assistance. Although in such cases timely communication, information gathering and coordination are crucial to assist those in need, the availability, proper functioning and supporting infrastructure of mobile technologies is still fraught with fragilities and limitations that prevent them from being fully effective (Patricelli et al. 2009; Townsend & Moss 2005; West & Valentini 2013; Guo & Su 2012; Effros et al. 2010; The Telegraph2011; Napolitano 2009; Xia 2011; Palen & Liu 2007; Yang et al. 2009; Zook et al. 2010). Here we discuss ways forward to overcome these limitations, proposing the incorporation of features to increase their resilience and effectiveness as mitigation tools at relatively low cost. Some of the measures proposed are not tied to daunting technological demands, but simply require unlocking capabilities already available, and affordably. Although our focus is mainly on natural disasters and global health crises, the solutions we identify are also useful to manage a wide range of emergencies that routinely affect millions of people around the world, especially in vulnerable and highly populated regions.

The Potential of Mobile Phones as Effective Aid Tools

We consider two major mobile phone categories: cell phones and smartphones. In addition to voice communication and text messaging capabilities, cell phones can include features such as digital cameras and radio receivers. Smartphones share these features, but have also become powerful computing machines that can handle complex tasks and accommodate a wider array of human-centric sensor devices (Srivastava et al. 2012). Those capabilities are increasingly explored through the development of software applications across a vast array of themes, including those useful for emergency preparedness and response (West & Valentini 2013). However, cell phones are cheaper, have more energy autonomy, and are still the device of choice among developing nations and regions, and more vulnerable populations (like the elderly), so they should not be overlooked for planning and preparedness purposes. Moreover, while some smartphone apps are also useful off-line, they still require access to a data network to be installed or run, hindering their usefulness when data transfer is not possible. Other communication devices such as tablets, computers and land-lines share several of the capabilities mentioned. However, the ubiquity of mobile phones at all times makes them uniquely suited as a critical resource for assistance in unexpected and disruptive scenarios, hence we will focus mainly on these devices.

We first briefly consider major fronts where current capacities and improvements in mobile phones can harness the power of rapidly developing technologies to serve society, especially in the fields of human health, environment and humanitarian assistance, alerts and emergencies.

Distress Calls and Establishing First Line of Contact

Mobile phones already represent the main means to establish the first line of contact to ask (or offer) help, and to obtain information through which help can be offered and achieved, and about the situation of relatives and friends. This is, by far, the most important and useful service that mobile devices offer under dire scenarios. Accordingly, most of the suggestions we will discuss are aimed to secure such capabilities under a wide range of stressing conditions.

Receiving Critical Information

When other means of communication (TV, radio, computers) are not accessible (e.g., out of reach or impaired), mobile devices can be the only source of information on evacuation plans and procedures, on the availability of shelters and other facilities, services and resources, weather forecasts, survival recommendations and other arguably lifesaving information.

Disseminating Data and Information to Aid Agencies

During earthquakes, hurricanes, floods, pandemics and other severe disease outbreaks, economic, societal and personal infrastructures are severely affected and governmental capabilities become quickly saturated due to the sheer demand for assistance? In such contexts timely communication, information gathering and coordination are crucial to assist those in need. Mobile phones can provide information and coordination of help to victims and vulnerable individuals, enable rapid data collection (Lurie et al. 2013) and even track in real-time population movements in the context of relief efforts (Bengtsson et al. 2011).

Other Applications

Other potential uses may range from those of psychological nature (as they can comfort and even entertain victims, helping lower panic and distress levels) to support of on-site data gathering and parallel computing processing just to name a few. For example, crowd sourcing to collect critically important information about a given situation or on a sustained basis to develop long-term records about environmental, demographic and health factors is also being considered by a wide range of organizations at the national and international levels.

These applications are intended to serve only as examples of practical and innovative ways that mobile devices can be used in service to society, and should not be viewed as an exhaustive list. In the next section we identify some of the current fragilities of mobile devices that limit their use for such conditions and applications, and propose the use of existing and emerging technologies to make them more resilient and useful in the future.

Existing Fragilities and Proposed Solutions

Critical capabilities are needed to ensure that mobile phones can serve at least the most essential roles previously discussed. Among obvious features, physical durability, energy autonomy, and network availability are most essential. But other features such as redundancy of frequency channels for communication are also important to turn these devices into effective aid tools in emergencies. We highlight current capabilities and limitations, and propose solutions based on existing and emerging technologies.

Off-line Operability

Both communications and access to remotely stored information are interrupted when the network structure of mobile operators is damaged. In such cases, off-line features can provide assistance.

One such important off-line feature has unnecessarily vanished, or is deactivated, from many modern mobile phones: the radio receptor. Although news, music and information can now be streamed directly from data networks through smartphone apps, direct access to radio stations should not be considered redundant. Data-streaming requires that several antennae and terrestrial structures provided by mobile operators function properly, whereas the signal from each radio station is broadcasted independently. Only one radio station suffices for the mobile user to make use of it. Radio stations have been shown to be important in crises (Xia 2011; Covello & Hyer2007; Baze 2012; National Association Broadcasters (NAB) 2013), and when combined with social media (if accessible) their impact can be largely amplified (BBC Media Action 2012). The National Association of Broadcasters states that the top 10 phones in the marketplace already include radio chips that are simply not activated (Murray 2010). Given the ease of reactivating radio receptors into mobile devices, their inclusion or activation in all future devices would be highly advisable. Moreover, because the costs of keeping an already existing feature would be likely very low, educating consumers about its effectiveness in emergency and disruptive scenarios could also contribute to a greater market penetration of these devices.

Another off-line feature that can be of great value in a disaster is geolocation capability. GPS access signals directly from satellites, hence does not necessarily depend on internet and data services. However, maps currently in use often access data in real time from on-line sources, rendering the information provided by the GPS nearly useless if map data cannot be downloaded. Users should thus always have a back-up map of their homes or destination and surroundings. Some smartphone apps that can work off-line that do not rely on or require a data network to run can be also useful (West & Valentini 2013).

Energy Autonomy

The short battery life of most smartphones (Neild 2013) is a major limitation in emergency situations. Electrical grids can be severely damaged following a disaster, and although a few hours of autonomy may suffice for emitting the first distress calls, most emergencies last much longer. In life-threatening circumstances (such as when victims are under the rubble) the interruption of service can be fatal or cause major damage (Daily Mail Reporter 2011).

Cell phones often have more autonomy, usually several days, as they lack the power-hungry screens and processors of smartphones. Indeed, they have already proven useful during power outages or breakdowns in developing countries (Bengtsson et al. 2011), and are a suggested back-up means of communication even for smartphone users (Stout & Neild 2013). While this may be wise advice, it is unlikely that most smartphone users will constantly carry two devices.

In addition to improving energy autonomy, creative methods of charging mobile phones should be considered. Solar cells, physical motion chargers, micro-fuel cells and external batteries can be purchased separately to provide greater autonomy. Interestingly, solar cells based on nano- and thin-film technologies that are embedded in or integrated in mobile phones are promising technologies. For example, solar powered phones were distributed in the aftermath of the catastrophic earthquake in the Chinese province of Sichuan during efforts to recover mobile communication (Guo & Su 2012; Yang et al. 2009). Some of those solutions could be integrated into future phones.

Alternative Channels of Communication

Where fiber optic cables and electricity are still operational (no doubt an optimistic scenario in a major catastrophe) the Internet or other packet-switched networks could still be used (Murray2010). But if mobile phone networks are down or overloaded, alternative channels of communication that are independent of external infrastructure (towers, cables) should be considered. One example is the critical role played by amateur and FRS/GMRS radios (Baze 2012). Ad-hoc networks by which each mobile phone in range functions as a relay point for other mobile devices also represent a promising technological avenue (Effros et al. 2010; Australian Broadcast Corporation ABC 2010). Another possibility is the integration of two-way radio channels into mobile devices, opening opportunities for close range communication (Zhou et al. 2012). There are also existing satellite phones which do not depend on the terrestrial infrastructure (Baze 2012). Indeed, those were the only devices that could be operated by medical relief teams in the aftermath of the Great East Japan Earthquake in 2011 (Fuse 2011). The ability to access space-based telecommunication networks directly, and affordably, from mobile phones would greatly increase resilience to disasters.

Physical Durability

Durability and ruggedness is not a common feature of most mobile phones. A quick glance at footages of natural disasters suggests that most currently available devices would likely become inoperable under the physical stress imposed during such events. In what is perhaps the early rise of a welcoming trend, some devices are under development with sturdier features such as shock resistance and water proofing (CNET 2013). The technology to improve the ruggedness of mobile phones, for example by using carbon fiber materials, is already available, and manufacturing costs do not seem to be a major barrier.

Distress Mode

Individuals who need to let others know about their whereabouts could greatly benefit from the activation of a distress mode in their devices. While saving as much energy as possible (e.g. by turning off non-essential features), a distress mode could regularly send pulse signals, which could be composed of a sound, a flash, and a radio signal containing data with personal information of medical relevance and individual location. This feature could be based on existing Personal Locator Beacons, SPOT Personal Trackers and the Iridium satellite system (Baze 2012) and could be built based on Digital Distress Identifiers that are already used by medically impaired costumers and as personal emergency alerts in accidents, as well as personal, community, state and regional emergencies. Even if victims fell unconscious, this feature (which could be built-in to enable its activation manually, by a voice command or even remotely by another device using a secure protocol) would increase the possibilities of a successful rescue.

Network Resilience

Although mobile phone networks are relatively resilient to external shocks, disasters can affect power supply, destroy towers, and cause complete loss of functionality (Townsend & Moss 2005; West & Valentini 2013; Bengtsson et al. 2011; Bengtsson et al. 2011; Baze 2012). During the superstorm Sandy in 2012, which took place in one of the most developed and prepared places on Earth, mobile phone networks were frequently down (Allen 2012; The Economist 2012). Network data links should therefore be not only more resistant to damage, but also redundant, diversified and able to activate back-up systems when antennas and cables are damaged (Patricelli et al.2009; Xia 2011; Fuse 2011). There is not much that consumers or manufacturers can do in this regard (other than promote redundancy by allowing mobile phones to use more than one network provider in the same device). Ultimately, the resilience of the communications network is a matter of national security. It is thus up to governments to detect fragilities in the system and overcome them by appropriate investments and implementation of a legal framework through which this can be achieved.

CONCLUSIONS

Most people live their entire existence without direct experience of major collective disasters. This shapes our perceptions as consumers and the features we seek in the products we purchase, which in turn guide the development and design of those products. But major public emergencies do happen and are in fact frequent worldwide (Lurie et al. 2013). Although mobile communication devices have already proven to be extraordinary aid tools in these situations (The Economist2009a; BBC Media Action 2012; The Economist 2007; West & Valentini 2013; The Telegraph 2011; Chen 2010), much more can be done to minimize their fragilities and increase their effectiveness. We believe that such concerns should be taken into account by consumers, policy makers, designers and manufacturing companies.

The features that make a mobile phone commercially attractive to daily use are not necessarily incompatible with those that make it useful in case of emergencies. For example, the improvement of battery life and physical resistance are not only critical in collective emergencies, they may also eliminate daily nuisances. This may be particularly important in developing and low income settings, where over 80% of the world population lives. In such cases, features such as robustness and battery life are expected to be even more crucial (e.g., if access to electricity is costly or unreliable). For example, a study among Bangladeshi students found that battery life was more important in purchasing decisions than the presence of a camera, color display, sound performance, memory capacity, style and even price (Siddique et al. 2013).

Other features, including the introduction of distress radar emission signals or walk-talk capabilities, could nevertheless increase costs. In such cases, consumers can be informed on the potential uses of such features, so they can create the demand for the market to fulfill.

The emerging field of Mobile Health (m-Health) focuses predominantly on software and application development, but largely takes hardware specifications for granted. But improvement of key features can make a difference in critical contexts, as even simple voice or text communication is not possible when devices – or their supporting infrastructure – are impaired. Important features can be incorporated by policy or by educating consumers (hence manufacturing companies) on the specifications that can be helpful in emergency conditions. With few exceptions (such as ensuring that all devices can receive radio signals), we favor the market-based approach, as regulatory measures quite often can dampen competition and innovation, and may hamper what is perhaps the most useful feature of mobile phones in a disaster: their ubiquity and vast penetration.

Companies and individuals who made the mobile revolution possible are the embodiment of how technological innovation and creativity can speed up the achievement of well-being for billons of individuals. In fact, the widespread use of the technological wonders represented by mobile phones may prove to be one of the greatest allies in the management of natural disasters, public health emergencies and humanitarian crises. Part of this extraordinary potential has already been achieved in currently available devices. However, even more would be possible if we can ensure robust communication for all those trapped in all too common "unthinkable" situations.

AUTHOR CONTRIBUTIONS

WJA conceived the essay and wrote the first draft. WJA, CSP and GRA contributed to the writing and with ideas, revised and approved the final version.

ACKNOWLEDGMENTS

We thank Lewis Kim, Ellis McKenzie, Bing Kung, Eduardo Beltrame and Alicia A. Levinsky (National Institutes of Health Library) for their valuable comments on manuscript. This research has been made possible by the support of the Fogarty International Center, National Institutes of Health.

Responsible editor: Thomas Krafft.

REFERENCES

1. Allen E (2012) First an Electricity Blackout and now Cell Phone Coverage is Down as Users in Manhattan Battle Signal Failures. Mail Online. accessed Nov 4, 2013 at http://www.dailymail.co.uk/news/article-2225217/Superstorm-Sandy-New-York-CELL-PHONE-coverage-users-battle-signal-failures.html
2. Australian Broadcast Corporation ABC (2010) Mobile invention could be desert lifeline. ABC. accessed Nov 4, 2013 athttp://www.abc.net.au/science/articles/2010/07/12/2951206.htm#.Ucxwlz7wKrh
3. Baze A (2012) Personal Emergency Communications: Staying in Touch Post-Disaster: Technology, Gear and Planning. Max Publications.
4. BBC (2013) Mobile phone hits 40th birthday. BBC. accessed Nov 4, 2013 at http://www.bbc.co.uk/news/technology-22013228
5. BBC Media Action (2012) Still left in the dark? How people in emergencies use communication to survive – and how humanitarian agencies can help. BBC Media Action. accessed Nov 4, 2013 athttp://downloads.bbc.co.uk/mediaaction/policybriefing/bbc_media_action_still_left_in_the_dark_policy_briefing.pdf
6. Bengtsson L, Lu X, Thorson A, Garfield R, von Schreeb J (2011) Improved response to disasters and outbreaks by tracking population movements with mobile phone network data: a post-earthquake geospatial study in Haiti. PLoS Med 8:e1001083

7. Chen BX (2010) Man Buried in Haiti Rubble Uses Iphone to Treat Wounds. Wired.com Gadget Lab. accessed Nov 4, 2013 at http://www.wired.com/gadgetlab/2010/01/haiti-survivor-iphone/
8. CNET (2013) Rugged phones posts on CNET. CNET. accessed Nov 4, 2013 at http://news.cnet.com/8300-5_3-0.html?keyword=rugged+phones
9. Covello VT, Hyer RN (2007) Effective Media Communication During Public Health Emergencies. A WHO Handbook. World Health Organization, Geneva. accessed Nov 4 athttp://www.who.int/csr/resources/publications/WHO%20MEDIA%20HANDBOOK.pdf?ua=1
10. Daily Mail Reporter (2011) "I"m not going to give up': Woman's last message to rescuers from beneath Christchurch earthquake rubble before her mobile phone battery runs out. Mail Online. accessed Nov 4, 2013 at http://www.dailymail.co.uk/news/article-1359354/New-Zealand-earthquake-A-mothers-message-beneath-Christchurch-rubble.html
11. Effros M, Goldsmith A, Médard M (2010) The rise of instant wireless networks. Sci Am 302:72-77
12. Fuse A (2011) Medical relief activities conducted by Nippon medical school in the acute phase of the great east Japan earthquake. J Nippon Med Sch 78:397-400
13. Guha-Sapir D, Santos I, Borde A (2013) The Economic Impacts of Natural Disasters. Oxford University Press, Oxford; New York.
14. Guo Y, Su XM (2012) Mobile device-based reporting system for Sichuan earthquake-affected areas infectious disease reporting in China. Biomed Environ Sci 25:724-729
15. Kenett DY, Portugali J (2012) Population movement under extreme events. PNAS 109:11472-11473
16. Lester R, Karanja S (2008) Mobile phones: exceptional tools for HIV/AIDS, health, and crisis management. Lancet Inf Dis 8:738-739
17. Lurie N, Manolio T, Patterson AP, Collins F, Frieden T (2013) Research as a part of public health emergency response. N England J Med 368:1251-1255
18. Murray MJ (2010) Communicating during a disaster. Anesth Analg 110:657-658

19. Napolitano D (2009) Getting the Message. Occup Health Safety 78:28-29
20. National Association Broadcasters (NAB) (2013) Equipping mobile phones with broadcast radio capability for emergency preparedness. NAB. accessed Nov 4, 2013 at http://www.nab.org/advocacy/issue.asp?id=2354&issueid=1082
21. Neild B (2013) Smartphone Power StrugglesWill we Ever Have Battery-Free Mobiles?. CNN. accessed Nov 4, 2013 at http://www.cnn.com/2013/02/27/tech/battery-free-mobile-phones/index.html
22. Palen L, Liu SB (2007) Conference on Human Factors in Computing Systems. In Proceedings of the SIGCHI, New York, NY. pp 727-736
23. Patricelli F, Beakley JE, Carnevale A, Tarabochia M, Von Lubitz DKJE (2009) Disaster management and mitigation: the telecommunications infrastructure. Disasters 33:23-37
24. Quadir I (2005) Mobiles Fight Poverty | Video on TED.com. accessed Nov 4, 2013 at http://www.ted.com/talks/iqbal_quadir_says_mobiles_fight_poverty.html
25. Siddique ZR, Jamil AA, Ali B (2013) Product features affecting buying decision for mobile phone handset: a study on tertiary students segment in Bangladesh. Eur J Bus Manag 5:139-146
26. Srivastava M, Abdelzaher T, Szymanski B (2012) Human-centric sensing. Philos Transact A Math Phys Eng Sci 370:176-197
27. Stout KL, Neild B (2013) Launch of Nokia's $20 Phone Begs Question: Is one Mobile Enough?. CNN. accessed Nov 4, 2013 at http://www.cnn.com/2013/02/25/tech/nokia-budget-phone-elop/index.html
28. The Economist (2007) Dealing with disasters: Flood, famine and mobile phones. The Economist. Accessed Nov 4, 2013 at http://www.economist.com/node/9546242
29. The Economist (2009) Eureka moments. accessed Nov 4, 2013 at http://www.economist.com/node/14483872
30. The Economist (2009) Mobile marvels. accessed Nov 4, 2013 at http://www.economist.com/node/14483896
31. The Economist (2012) Success and failure after the storm. The Economist. accessed Nov 4, 2013 at http://www.economist.com/blogs/schumpeter/2012/11/sandy

32. The Telegraph (2011) Turkey earthquake: Four pulled alive from rubble after victim calls for help on mobileTelegraph.co.uk. Accessed Nov 4, 2013 athttp://www.telegraph.co.uk/news/worldnews/europe/turkey/8846337/Turkey-earthquake-Four-pulled-alive-from-rubble-after-victim-calls-for-help-on-mobile.html
33. The World Bank (2012) Information and Communications for Development 2012. Maximizing Mobile. The World Bank, Washington, DC. accessed Nov 4, 2013 at http://www.worldbank.org/ict/IC4D2012
34. Townsend AM, Moss ML (2005) Telecommunication infrastructure in disasters: preparing cities for crisis communication. accessed Nov 4, 2013 at http://www.nyu.edu/ccpr/pubs/NYU-DisasterCommunications1-Final.pdf
35. West DM, Valentini E (2013) How Mobile Devices are Transforming Disaster Relief and Public Safety. Brookings Institution, Washington DC. accessed Nov 4, 2013 at http://www.brookings.edu/research/papers/2013/07/16-mobile-technology-disaster-relief-west
36. Xia Y (2011) Mobile Communication. In: Katz J (ed) Dimensions of Social Policy, NJ: Transaction Publishers. pp 87-102
37. Yang C, Yang J, Luo X, Gong P (2009) Use of mobile phones in an emergency reporting system for infectious disease surveillance after the Sichuan earthquake in China. Bull World Health Org 87:619-62
38. Zhou A, Shi L, Mao Y, Tang J, Zeng Y (2012) Diffusion of new technology, health services and information after a crisis: a focus group study of the Sichuan "5.12" earthquake. Int J Health Plann Manage. doi:10.1002/hpm.2137
39. Zook M, Graham M, Shelton T, Gorman S (2010) Volunteered geographic information and crowdsourcing disaster relief: a case study of the Haitian earthquake. World Medical Health Pol 2:7-33

Citations

CHAPTER 1

Stephanie M. Fletcher, Jodi Thiessen, Anna Gero, Michele Rumsey, Natasha Kuruppu, and Juliet Willetts, "Traditional Coping Strategies and Disaster Response: Examples from the South Pacific Region," Journal of Environmental and Public Health, vol. 2013, Article ID 264503, 9 pages, 2013. doi:10.1155/2013/264503.

CHAPTER 2

Yoshihiko Hasegawa, Akira Sugai, Yosuke Hayashi, Yuta Hayashi, Shoji Saito, and Toshiki Shimbori, Improvements of Volcanic ash Fall Forecasts Issued by the Japan Meteorological Agency, doi:10.1186/s13617-014-0018-2.

CHAPTER 3

Emily Wilkinson, Beyond the volcanic crisis: co-governance of risk in Montserrat, doi:10.1186/s13617-014-0021-7.

CHAPTER 4

Max T Kinateder, Erica D Kuligowski, Paul A Reneke, and Richard D Peacock, Risk Perception in Fire Evacuation Behavior Revisited: Definitions, Related Concepts, and Empirical Evidence, doi:10.1186/s40038-014-0005-z.

CHAPTER 5

Soichiro Suzuki, Satoshi Hasegawa, and Masayuki Okugawa, Remote Control System of Disaster Response Robot with Passive Sub-Crawlers Considering Falling Down Avoidance, doi:10.1186/s40648-014-0020-9.

CHAPTER 6

Francis EO Opiyo, Oliver V Wasonga, and Moses M Nyangito, Measuring Household Vulnerability to Climate-induced stresses in pastoral rangelands of Kenya: Implications for resilience programming, doi:10.1186/s13570-014-0010-9.

CHAPTER 7

Wladimir J Alonso, Cynthia Schuck-Paim, and Ghassem R Asrar, Global health and natural disaster alerts: preparing mobile phones to endure the unthinkable, doi:10.1186/2194-6434-1-24.

Index